Food Systems and Agrarian Change

Edited by Frederick H. Buttel, Billie R. DeWalt,
and Per Pinstrup-Andersen

Research and Productivity in Asian Agriculture
by Robert E. Evenson and Carl E. Pray

Searching for Rural Development:
Labor Migration and Employment in Rural Mexico
by Merilee S. Grindle

Diversity, Farmer Knowledge, and Sustainability
edited by Joyce Lewinger Moock and Robert E. Rhoades

Networking in International Agricultural Research
by Donald L. Plucknett, Nigel J. H. Smith, and Selcuk Ozgediz

Agriculture and the State: Growth, Employment,
and Poverty in Developing Countries
edited by C. Peter Timmer

Transforming Agriculture in Taiwan: The Experience of
the Joint Commission on Rural Reconstruction
by Joseph A. Yager

DIVERSITY, FARMER KNOWLEDGE, AND SUSTAINABILITY

EDITED BY
Joyce Lewinger Moock
and Robert E. Rhoades

Cornell University Press

ITHACA AND LONDON

International Standard Book Number 0-8014-2682-0 (cloth)
International Standard Book Number 0-8014-9968-2 (paper)
Library of Congress Catalog Card Number 92-52768
Printed in the United States of America
Librarians: Library of Congress cataloging information appears on the last page of the book.

Contents

Foreword

The last decade of the twentieth century will be a challenging and exciting period for international agricultural research. Scientists are only beginning to glimpse the revolutionary technical breakthroughs in genetic engineering and biological pest control that will help feed an increasingly hungry world. Simultaneously, we have become painfully aware of the negative pressure that the growing global population is placing on the earth's natural resources. This concern is encapsulated in the theme of sustainability, a still vague but increasingly powerful force in remaking our agricultural research agendas.

This book argues that, along with biological science, past and present indigenous farmer knowledge can play a key role in sustainability. The authors of the chapters—many of whom are Rockefeller Foundation Social Science Fellows, from both developed and developing countries—all work closely with plant breeders and crop production scientists. By tackling the links between farmer knowledge, technology development, and sustainability, they have carried their research skills into largely uncharted waters. Their research empirically demonstrates that farmers are active experimenters, plant experts, and local soil scientists in their own right. These findings, in turn, illustrate the practical feasibility of incorporating farmers' knowledge into the design of sustainable food systems. Above all, as applied social scientists, the authors have shown how the reality at the farm level—in fields, households, and markets—is important to research priorities of nationl and international organizations. No matter how much science and technology investment is made in laboratories or experiment stations, payoffs must ultimately come through the effectiveness and sustainability of local food systems, particularly those located in marginal environments where infusion of new technologies has been slow. Future science and farmer sciences, as this volume demonstrates, can be complementary in the coming struggle to develop self-sufficient and sustainable livelihoods for all peoples of the human family.

RICHARD L. SAWYER
Director General Emeritus
International Potato Center
Lima, Peru

Acknowledgments

The Rockefeller Foundation and the International Potato Center (CIP), Lima, Peru, supported the Farmers and Food Systems Conference from which the papers in this volume are drawn; the Rockefeller Foundation also provided funding for the preparation of the book. Margaret Novicki served as a consultant editor and stimulated a high standard of respect among our social science authors for the English language. Adrianne Nagy, MaryAnn Knotts, and Heather Bent Tamir oversaw the numerous practical details of the preparation of the manuscript in conversations with the authors and Cornell University Press.

The editors gratefully acknowledge the efforts of CIP staff in organizing the 1988 meeting in Lima, and especially the hospitality and insightful contributions of the Peruvian national agricultural scientists, who participated in the week-long discussions.

J.L.M. and R.E.R.

Contributors

AKINWUMI A. ADESINA, an economist formerly with the International Crops Research Institute for the Semi-Arid Tropics (ICRISAT), is a research scientist at the West Africa Rice Development Association (WARDA).

STEPHEN BIGGS is a senior lecturer in the School of Developmental Studies, East Anglia University.

R. JAMES BINGEN is an associate professor at Michigan State University.

KAREN ANN DVOŘÁK, an economist formerly with the International Fertilizer Development Center (IFDC) and the International Crops Research Institute for the Semi-Arid Tropics (ICRISAT), is a research scientist at the International Institute of Tropical Agriculture (IITA).

S. K. EHUI, an economist formerly with the International Institute of Tropical Agriculture (IITA), is a research scientist at the International Livestock Center for Africa (ILCA).

PETER EWELL is a regional social scientist with the International Potato Center (CIP).

PABLO EYZAGUIRRE is an anthropologist with the International Service for National Agricultural Research (ISNAR).

SAM FUJISAKA is an agricultural anthropologist with the International Rice Research Institute (IRRI).

ABE GOLDMAN, formerly with the International Institute of Tropical Agriculture (IITA), is an assistant professor of geography at the University of Florida.

xi

ROBERT W. HERDT is the director of the Agricultural Sciences Division of the Rockefeller Foundation.

B. T. KANG is a research scientist with the International Institute of Tropical Agriculture (IITA).

JOHN K. LYNAM, an economist formerly with the International Center for Tropical Agriculture (CIAT), is a field staff member of the Rockefeller Foundation.

JEAN MCALLISTER, formerly a research assistant at the International Service for National Agricultural Research (ISNAR), is a graduate student in the Department of Anthropology, Columbia University.

DEBORAH MERRILL-SANDS is a senior research officer at the International Service for National Agricultural Research (ISNAR).

JOYCE LEWINGER MOOCK is the associate vice president of the Rockefeller Foundation.

SUSAN POATS is a social scientist in the Cassava Program at the International Center for Tropical Agriculture (CIAT).

GORDON PRAIN, an anthropologist formerly with the International Potato Center (CIP), coordinates the UPWARD Network in the Philippines.

ROBERT E. RHOADES, formerly a social scientist with the International Potato Center (CIP), is the head of the Department of Anthropology, University of Georgia.

DEBORAH S. RUBIN, an anthropologist formerly with the International Food Policy Research Institute (IFPRI) and the International Centre for Insect Physiology and Ecology (ICIPE), is an assistant professor of anthropology at the University of the Pacific.

URS SCHEIDEGGER is a seed production specialist with the INIAA-CIP-COTESU Seed Project.

D. S. C. SPENCER is the director of research at the International Institute of Tropical Agriculture (IITA).

LOUISE SPERLING is an anthropologist with the International Center for Tropical Agriculture (CIAT).

FULGENCIO URIBE was formerly an agronomy assistant with the INIAA-CIP-COTESU Seed Project.

JOACHIM VOSS, an anthropologist formerly with the International Center for Tropical Agriculture (CIAT), is director of the Sustainable Production Systems Program of the International Development Research Centre.

TIMOTHY OLALEKAN WILLIAMS is an economist with the International Livestock Centre for Africa (ILCA).

Diversity, Farmer Knowledge,
and Sustainability

Introduction

Joyce Lewinger Moock

The demand for food in developing countries during the first quarter of the twenty-first century will be staggering. By the year 2025, the world's population will have exploded to 8.5 billion inhabitants, 83 percent of whom will live in Third World nations. More food will need to be produced within the lifetime of that future generation of farmers than has been grown in the entire 12,000 years since the Neolithic revolution. Even with the miracles of biotechnology on the horizon, food production alone will not avert chronic food deficits for a large proportion of the population, particularly the poor living in fragile and harsh rainfed environments. For them and for millions of others, the agricultural sector will be expected to play a crucial role in the generation of income and employment as a means of ensuring food security. Agricultural planning for the year 2025 will also take place within the context of escalating pressures on the natural resource base and their dire implications for the long-term sustainability of food production systems.

The combination of rapid population growth, poverty, and deterioration of global natural resources represents a powerful challenge for the current capabilities of the international establishment that organizes agricultural research on a global basis. This establishment, consisting of the thirteen publicly funded institutes of the Consultative Group on International Agricultural Research (CGIAR)[1] and as-

[1] The Consultative Group on International Agricultural Research is sponsored by the World Bank, the United Nations Development Program, and the Food and Agriculture Organization, and is headed by a senior official of the World Bank. In 1990, its thirty-eight donor

sociated research centers, attempts to fill gaps in research beyond the current capabilities and often the immediate interests of national programs. At its inception in 1971, CGIAR had tightly focused and measurable goals: increasing the production of poor people's food staples and strengthening national programs through training and research dissemination. The success of the Green Revolution (a term used for the explosive increases in wheat and rice production in developing countries brought about by improved varieties and the use of chemical fertilizers and irrigation) justified the narrow technological focus and, along with it, the boundaries of legitimate problem-solving approaches.

Over the past twenty years, the intellectual horizons of the international agricultural research system have clearly broadened. Recognition of the multitude of farming systems and institutional and policy environments unfavorable to Green Revolution approaches, along with rising alarm over the erosion of natural resources, has provided a context demanding new criteria for the assessment of viable technology. Added to the traditional tests of productivity and stability are now such considerations as diversification of food products for nutritional security, potential for income and employment generation, improvements in the welfare of rural women, and, most recently, conservation of natural resources and the sustainability of production systems over time. Together, they illustrate the trend in agricultural research to move analysis of the human factor in technology development to the forefront of scientific concerns.

Yet, the complexity and immense variety of farming systems in the less favorable environments of Asia, Africa, and Latin America remains poorly understood. Biological and physical scientists have found it easier and of greater short-term benefit to deal with regions characterized by relatively homogeneous soils and moisture levels, domination of a single crop, possibilities for adoption of a new crop over a wide area, and easy adjustment in input supplies than to focus on the problems of marginal environments. The latter often are characterized by foods grown under rainfed conditions, ecological systems with edaphic stresses, political districting units with diverse farm structures and food demands, poorly developed input delivery systems, and crops with severe postharvest constraints.[2] Social scientists,

---

members pledged $234 million in core support of the thirteen international agricultural research centers and their associated institutes.

[2] See John Lynam, "On the Design of Commodity Research Programs in the International

in contrast, have conducted a great deal of research in the less favorable regions, but these investigations have not, in general, been effectively linked with the issue of basic food production or technology development and deployment. Despite recent advances in the formation of interdisciplinary teams, most technical research targeted for smallholder units continues today with little attention to the characteristics of farming populations and their resources and strategies, and most social science research on rural transformation is devoid of technological comprehension.

This volume presents a set of reflective commentaries on the complexity of new social criteria for technology assessment and generation, and the challenges they pose for agricultural research. Many of the chapters are concerned with the role of farmer knowledge and decision-making in relation to technological options. All focus on the ways in which social and economic perspectives can interact with and fortify the biologically based work of the international agricultural research establishment.

The authors bring to their task an unusual perspective. They have served as members of integrated social science and biological research teams within the international agricultural research institutes. All lead authors, with one exception, are current fellows or alumni of the Rockefeller Foundation Social Science Fellowship Program in Agriculture, which places exceptional young scholars with CGIAR and associate centers for two-year field assignments.[3] The purpose of the program is to assist the centers in (a) developing better-calibrated mechanisms for problem identification; (b) formulating viable, clearly focused research strategies, usually for commodity programs, in terms of complex socioeconomic goals; and thus (c) translating social perspectives into tangible technological outcomes. This is a far different and more difficult role for social scientists in the international centers

---

Centers," in David Groenfeldt and Joyce Lewinger Moock, eds., *Social Science Perspectives on Managing Agricultural Technology* (Sri Lanka: International Irrigation Management Institute, 1989).

[3] This program was initiated in 1974 as a response to the lack of training mechanisms to prepare future generations of first-rate North American social scientists for participation in the growing international agricultural research network. A component was added in 1985 to place top-caliber African nationals at international agricultural centers in Africa. The program, run as a cost-sharing operation with the international centers, has now made seventy-one awards, including ten to African nationals. Biennial conferences bring the fellows and selected alumni together to discuss aspects of biological or resource management research that have benefited from social analysis. The 1990 site was the International Institute of Tropical Agriculture in Nigeria.

than that assumed in the dominant perception: that they are concerned largely with monitoring predesigned technologies in terms of end-users' perspectives. The program is intended to incorporate social analysis higher upstream—that is, in the identification and operationalization of researchable problems.

The chapters in this volume were first presented in draft form at the Farmers and Food Systems workshop in late 1988 in Lima, Peru, sponsored jointly by the International Potato Center (CIP) and the Rockefeller Foundation. The central objectives of the meeting were to take stock of the evolving criteria for technology development and evaluation within the international agricultural research system, to identify conceptual and methodological constraints, and to assess the status of multidisciplinary scientific approaches to problem solving. Along with the fellows, representing fourteen international agricultural centers, participants included Rockefeller Foundation and CIP staff members and a number of Peruvian agricultural scientists.

Despite the diverse mandates (specific crops, livestock, environmental zones, food policy, institutional innovations) of the centers represented at the meeting, a set of common concerns for international agricultural research formed the major topics for debate:

- Are the new objectives of the global agricultural research establishment resulting in nontechnical research with little direct relevance to agricultural science and thus producing findings impossible to operationalize?
- To what extent should farmers' indigenous knowledge and knowledge acquisition processes influence the setting of national and international agricultural research agendas?
- How can agricultural research be structured so that farm-gate issues can be nested within a national and global food systems perspective involving such multiple policy issues as staple food production, exports and imports, expanding markets, land tenure, employment, and food self-sufficiency?

While the meeting did not resolve these questions, there was considerable value in attempting to frame them in a clear and coherent fashion. Much discussion concerned the trend among farming systems researchers to examine technical change from the perspective of a single variable—rural employment, food production, food consumption, food prices, land-use intensity, and others—rather than as the totality of several effects. Most participants felt that technology assessment criteria are value laden and, as such, must be investigated empirically from the framework of a well-defined farming unit and from a household decision-making perspective that involves farmer knowledge, the local market economy, and government policies. Participants also agreed that more research is needed to interpret diversity at the farm and community levels in order to anticipate changes in social organization and output specialization resulting from market development.

From the discussions in Lima, five central themes emerged, which are reflected in the chapters presented in this volume. The first, flagged by Pablo Eyzaguirre, stresses the problems of scale that result from the new global research agenda calling attention to issues of diversity, sustainability, and outreach to resource-poor farmers in unfavorable agroecological environments. During the past two decades, the division of labor between the international agricultural research centers and their national program clients compartmentalized generalizable scientific agricultural research from highly specific, empirical studies of heterogeneous farming systems. The new agenda, representing the next and more complex phase of work for the global agricultural research system, forces the international centers to cope with issues of immense diversity in agroecological conditions and farming practices at the basic research formulation stage. Thus, the nature of the agenda itself implies not only broadened intellectual horizons for CGIAR but also what may be a fundamental shift in its problem-solving strategies, bringing it much closer to the mission of national research systems in developing countries.

The new interest in sustainability is a case in point. The chapter by John K. Lynam and Robert W. Herdt addresses the question of how to conceptualize sustainability as a criterion for evaluating agricultural technologies and illustrates the difficulties of applying it to technology design. Sustainability, they argue, is relevant only when a system using a technology has been well specified over appropriate time and space dimensions, and therefore the criterion is difficult to apply empirically beyond the farming systems level. Furthermore, "unsustainability is often locally or regionally defined and depends on such factors as the rate of increase in exogenous demand on the system, agroclimatic environment, and the relative intensity (generally in land use) of existing systems. Targeting thus becomes a key issue for agricultural research programs where sustainability is a principal objective" (p. 211). What does this imply for the international agricultural research system, which is geared to generalizable research issues within a commodity focus and cannot accommodate those involving high levels of diversity?

Neither Eyzaguirre nor Lynam and Herdt attempt to resolve the ambiguities that characterize sustainability research for the international agricultural research establishment—how to organize research on problems whose solutions are location-specific and how to organize biological research when the focus is on the whole agricultural system and its productivity rather than on that of its individual components. Only experience can provide answers for these dilemmas. Both chapters, however, highlight the fact that dealing with sustainability and diversity may be a question not simply of grafting on new factors but of having an altogether different framework for analysis. Such a framework, in this case, involves an understanding of subsystem interactions, farmers' evaluation criteria, and the processes by which farming families adjust to a changing external environment.

The workshop's second theme highlights the linkage between farmer knowledge and world science. Farmers possess a vast store of experiential knowledge about little-known cultivars and plant species, variations in the soil quality of their farms, the climatic conditions they face, disease and local pests, and management of the natural resource base—knowledge far richer than the research team can acquire on intermittent visits. The chapters by Joachim Voss, Gordon Prain et al., Sam Fujisaka, Karen Ann Dvořák, and Louise Sperling show that farmers experiment with many farming and conservation techniques, emphasize diversity in their production systems as a source of both innovation and insurance against failure, and evaluate technology on the basis of criteria that may differ widely in their totality and internal weights from criteria used by national and international scientists.

But while the authors of this volume stress the value of tapping indigenous farmer knowledge of production, management, and the environment as a way of opening up new areas of inquiry on diversity and sustainability, concerns are expressed about the limitations of local folk science and of the private criteria farmers use in technology assessment. Several authors point out that farming systems are often well adapted to the climatic, biological, and economic conditions of specific areas, except in cases where rapid population growth, climate change, or market penetration occur so suddenly as to introduce ecological instability. The goal is not simply to strengthen the existing capacity of farmers to find their own technological solutions to a changing environment, but to provide new technological means with which they can regain control of the agricultural system. Simeon K. Ehui et al., Lynam and Herdt, and Timothy Olalekan Williams further note that even farmers concerned with sustainability tend to make production and management choices based on projections of family welfare in terms of current or near-term profits and costs.

Mobilizing farmers' experiential knowledge about their production base while at the same time broadening their comprehension of the full range of available technological opportunities emerging from world science is essential if sustainable agricultural growth is to occur. According to Voss and Lynam and Herdt, the translation of ecological priorities into adoptable technologies that rival agrochemicals and other deleterious environmental options will place heavy demands on farmers' knowledge about methods to improve soil quality, enhance genetic diversity, and control pests and disease through biological means. It therefore follows, as Eyzaguirre contends, that research itself rather than its output should be taken to the targeted environment.

The third theme of the workshop centers on ways to involve farmers directly in technology generation and evaluation. The ideal, agreed the meeting participants, is to forge full-fledged collaboration with farmers in applied research programs, particularly those focused on breeding, conservation, and the relief of principal management constraints. This volume provides descriptions of a number of workable arrangements for incorporating farmers' knowledge and experience early on in the research process. Most are creative variations on the "farmer-back-to-farmer" approach pioneered by the International Potato Center a decade ago, in which farmers work alongside the external research team in defining research priorities, identifying potential solutions, and testing the impact of the technologies generated. Sperling describes one especially interesting extension of the model in Rwanda, where local female bean experts are invited to the research station to critique varietal trials and select varieties to test in their own fields. The objective is to combine the farmers' knowledge with the breeders' talents to produce an adaptable product which is better than either could produce alone.

As for the results of such participatory experiments, Prain et al. present *la papa simpática* ("the friendly potato"); Voss and Sperling offer ecologically sound bean varietal mixtures; Fujisaka points to improved traditional farm implements and soil conservation techniques, among other innovations; and Dvořák highlights more flexible recommendations for fertilizer applications. Numerous other examples are offered in the emerging literature on participatory farmer research, a large portion of which is cited in this volume.

The progress of these approaches will depend on how well accompanying tensions can be addressed. Farmers' evaluations are slow and unpredictable. Farmers themselves differ in their criteria according to their economic status, market dependence, gender, and age. As Sperling notes, "Given the diversity of farmers' preferences and constraints, a tacit decision is made as to whose problems should be given priority" (p. 97). Farmer collaboration at the problem identification stage may move the research focus beyond a fixation with yield and even beyond agricultural production to overstep the bounds of the partner organization's research mandate. In this regard, the scientist may be left to reconcile different data sets—those based on conventional yield information and those based on farmers' measures of total productivity. Nonetheless, the chapters in this volume provide evidence that by involving farmers fully as partners, research programs will have a greater chance of resulting in technologies well adapted to local and national agricultural systems.

The need to produce an appropriate range of technological and agricultural development options as a response to diversity is the fourth theme to percolate from the meeting. Both Abe Goldman and Akinwumi A. Adesina argue that many future changes in farming systems will originate as adaptations or extensions of existing practices rather than as wholly novel transplants. Given the enormous variety of agroecological and social conditions in the less favorable areas, the need for greater flexibility and scope in the research programs of the international centers, along with a more decentralized approach in setting priorities, would seem intuitive. The chapters by Voss and Prain et al. indicate that farmers are looking not for an ideal variety but for an ideal range of varieties for diverse ecological conditions and uses. Voss further underscores the need for research strategies that will both "increase the quantity and improve the quality of the genetic diversity being conserved by farmers" (p. 35).

But the workshop also made clear that the development of an appropriate range of farming interventions goes beyond agronomic measures. Many questions were raised about the value of microlevel approaches that help farmers operate better under resource-poor conditions as opposed to a broader focus on changing those aspects of their conditions amenable to trade and institutional arrangements affecting land distribution and commodity and labor prices. The lack of attention to these latter issues was vigorously noted as a major shortcoming of the previous era of farming systems research.

Even the notion of diversity itself as a fixed characteristic of smallholder production systems came under attack. Lynam and Herdt note that population growth and resource constraints in many parts of Asia have led to market development and commodity specialization, thus lessening crop diversity at the farming systems level while maintaining it at the national or regional level. It may be possible, therefore, for increasing market dependence to reduce pressure on some agroecologies and enhance productivity of the overall system without changing the underlying production technology. Yet, the extent to which this thesis applies in the foreseeable future to sub-Saharan Africa, the region on which most of the chapters in this volume focus, is likely to be limited.

The fifth theme examines the nature of technology transfer and the role of the international agricultural research centers in the process. The workshop participants agreed that the job of the centers is not to transfer technology across agroecological lines (a difficult, if not impossible, task), but to transfer the means of creating productive agricultural technology. This includes, in addition to relevant substantive components of technology, the capacity for technology design, adaptation, and generation.[4] National systems that lack the capability for sound applied and adaptive research cannot be efficient borrowers or producers of agricultural technology. Nor can they anticipate the equity problems that may arise from technological innovations or—as Deborah Rubin's chapter implies—predict the complex adjustments that families and societies may need to make.

In meeting the needs of resource-poor farming households in complex and heterogeneous food and agricultural systems, the global agricultural research establishment faces its greatest challenges in terms of institutional constraints and biases. Location-specific research is expensive. Investigations of total productivity at the farm level involving multiple systems interactions and complex experimental designs face the pressures of funding cycles that do not permit adequate analysis of large quantities of data. The gleaning of farmer knowledge and incorporation of participant research and information-exchange approaches are time-consuming. In addition, as several of the authors note, the lack of indigenous scientists in national systems with the requisite training and experience to conduct multidisciplinary, highly detailed, problem-solving research is a severe impediment to developing the type of on-farm, broad-based, client-oriented investigative foci proposed. Thus, as the objectives of the international agricultural research establishment broaden and research strategies become more sophisticated, national research systems are likely to be burdened with

[4] See Yujiro Hayami and Vernon Ruttan, *Agricultural Development: An International Perspective* (Baltimore: Johns Hopkins University Press, 1985).

research tasks and new areas of accountability for which they are ill prepared.

The chapter by Deborah Merrill-Sands et al. provides a partial solution to this dilemma by suggesting practical approaches to research planning and implementation which complement and build on existing capacities. Such approaches attempt to link on-farm and experiment station research, on-farm research and farmers' evaluations, researchers working in different disciplines or on different commodities, and on-farm research and technology transfer agencies. However, the analytic reorientation necessary for scientists to seriously address the new criteria for technology assessment is extensive and should not be underestimated. National agricultural research appears in too many cases to be adapted more to the reality of weak research institutions than to the conditions of resource-poor farming systems. It therefore falls to the international centers to take the lead in defining the objectives and parameters of the task, contributing methodology for characterizing diversity and measuring sustainability, disseminating relevant information for policy choices, and providing training through selective on-site research.

The global agricultural research establishment has gradually moved from the focused goal of eliminating national food deficits to a concern with alleviating hunger and poverty and protecting the sustainability of the natural resource base. The numerous, and sometimes contradictory, objectives for which the international research system is now held responsible overwhelm the current arsenal of research tools and strategies. The chapters that follow wrestle with how the international centers and national systems can convert these commendable objectives into operational research programs covering very complex and heterogeneous agricultural environments in which millions of poor farm families live.

I

Farmer Knowledge, World Science,
and the Organization of
Agricultural Research Systems

Pablo Eyzaguirre

Introduction

The study of farming practices yields two major insights central to the management and sustainability of food production for resource-poor farmers. First, research within small-scale agrarian communities has highlighted the importance of diversity in sustainability and the management of farm resources. Second, field research has allowed us to glimpse the vast fund of technical knowledge that farmers possess about their farming systems. This chapter aims to link the insights derived from microlevel studies of farming communities to the agendas of national agricultural research systems and international agricultural research centers.

Linking the issues at the level of resource-poor farming communities to national or international research programs entails a consideration of the problems of scale. What is important to address at a microlevel may not be economic or even possible at the national or international level. Yet, as the global research agenda increasingly

The author thanks Joyce Lewinger Moock and Robert Rhoades for their review and comments on earlier drafts of this paper; Andrew Okello and Bonnie McClafferty assisted in the data collection and analysis. The Rockefeller Foundation provided initial funding for this research.

calls attention to the issues of diversity and sustainability, along with the need to extend the benefits of agricultural technology to resource-poor farmers in less favorable environments, the two levels are being brought closer together (Ruttan, 1988; Schuh, 1988; TAC, 1988).

Important questions arise as a consequence. Are the issues that social scientists and agriculturalists analyze in such detail at the micro-level of the farming community significant at the level of international agricultural research centers? Can and should the global agricultural research system address the issues of diversity and small-scale units and incorporate farmer knowledge into its agenda?

The answers to these questions are not exclusively in the domain of new agrotechnology. There is a danger that research systems may be promising too much (Schuh, 1988). This is especially true for many small countries. While their agricultural research needs are great, the absolute limits of size and scant resources may require specific strategies for national research.

Diversity and the Farming Community: Policy Implications

Numerous studies have emphasized the importance of diversity in small-scale tropical farming systems (Harwood, 1979; Merrill-Sands, 1986; Norman et al., 1982; Turner and Brush, 1987). An equally important dimension is the degree of change within farming systems. Studies at the level of farm households have documented the rapidity with which farmers change to exploit new niches or economic opportunities, provided they do not have to sacrifice their diverse base of production (Eyzaguirre, 1987; Moock, 1986). Farmers are often eager to innovate, as long as they do not have to put all their eggs in one basket. Diversity in their systems of production is both the source of innovation and the requisite insurance against failure.

At the level of national agricultural and agrarian policy, there is often a disjuncture in the approaches to diversity by small producers and by policymakers. This was the case in one small African country where the author conducted research on farming systems and the role of national policies in ecological and agrarian change. A diverse set of small-scale farming systems produced the bulk of the country's food and had potential to expand production of the principal export, cocoa (Eyzaguirre, 1987). National economic and agrarian policies, however, were directed only at the estate sector with its restricted scope of products and high costs of production. Despite the clear potential for food production and income generation inherent in the small-scale systems, this potential was ignored by national policy. Farmers regarded diversity as a source of both innovation and security, which they exploited and managed. National agricultural planning and policy on technological innovation, on the other hand, failed to take diversity into account. This was primarily for political and institutional reasons but also because large-scale, homogeneous units are easier to plan for and less subject to rapid change.

In general, a multiplicity in the types of farms and production systems makes agricultural sector planning and the delivery of services, inputs, and new technology more difficult. Farm-level research is therefore needed to interpret this diversity to formulate options for agricultural development and technological change. It has become increasingly apparent within both national systems and international centers that social science research should play a more important role in linking agricultural research systems and objectives to the national policy level (Elliott, 1990).

The Contribution of Farmer Knowledge to Research on Agricultural Diversity

A growing concern of the world scientific community is the promotion of research that fosters diversity and sustainability of the agricultural resource base. Tapping into the fund of indigenous technical knowledge can open up new fields of inquiry on these issues. Farmers may help to identify new crop lines and plants that can be the basis of more specialized, nontraditional, high-value export crops that are the comparative advantage of many small countries.

Research in two key areas could go some way toward solving these problems [the increasing specialization and uniformity in world food crops and the neglect of the traditional food staples of the rural poor]. First, agricultural scientists need to look more at subsistence crops or under-utilized plants that are the staples of the rural poor in the Third World. Secondly, the need to work toward the development of novel foods or industrial feedstocks in the developed countries . . . must rely on independent initiatives from the international agricultural research community. (de Groot, 1988)

Linking indigenous knowledge with world science can move the work of agricultural research in the direction of greater diversity and sustainability. International and national agricultural research systems can then extend their impact to reach the low-resource farmers in small-scale units on the margins of agricultural science and development.

Diversity and the International Agricultural Research Centers

The international agricultural research centers have arrived at a similar impasse with the concept of diversity. One response is to try to produce technological products for a wider range of environments. This places an unrealistic goal on the centers, whose comparative advantage is their ability to produce generalizable products rather than highly specific ones. Another approach has been to expect that the benefits from the centers' successful work on the major commodities produced by larger-scale agricultural regions (with great potential for increasing productivity with additional input) would spill over to marginal areas with small-scale units and more diverse systems.

In developing regions with comparatively less ecological diversity and large areas dominated by comparatively few crops, technological innovations in one or two crops have led to significant increases in food supplies and farm incomes. There are still many underdeveloped rural areas less touched by these advances in agrotechnology (TAC, 1988). This is partly because of less favorable environments in these areas, but also because the agroecological and agrarian units for the adaptation and adoption of new technology are small. Coping with the diverse conditions of farmers and their systems of production is a major problem for development-oriented agricultural science. The areas that remain to be addressed by the international agricultural research centers' programs of technological innovation are limited by the small scale of their respective social and political units.

The comparatively weak impact of agricultural research on African agriculture, for example, is due in large measure to Africa's agroecological, political, cultural, and social diversity, which reduces the scope for change and adoption of innovation. The mandates of the international centers identify national research systems as their major clients. Many of the developing countries in this client group that have yet to register significant growth in agricultural production and that adopt new technologies at low rates have less than five million people and proportionately large agrarian sectors.

There are forty-five countries in sub-Saharan Africa. Their average population is approximately six million, and twenty-five of them have populations of less than five million (Jahnke et al., 1987). At the local level, this set of factors produces numerous distinct farming systems on a small scale—within a village, or even a household—coupled with a rapid rate of change in these systems. For the international centers, the design and transfer of technology becomes more complicated when a targeted group of farmers and production systems within a broad agroecological zone occurs in more than one country. National agricultural policies, institutions, and infrastructures are crucial for the transfer and introduction of new technology. Effective national research systems must be in place, and socioeconomic research should play an important role in the transfer and adaptation of new technology.

Coping with Diversity and Scale: The Small-Country Problem

The inability to address Africa's diversity has made it difficult for the international centers and national research systems to develop appropriate technologies for the small farmers who make up the bulk of sub-Saharan Africa's population. The transfer and adoption of new technology is made more difficult by the many governments and smaller-scale societies that must provide the mechanisms for technology transfer, the policy environment, and infrastructure to support the adoption of new technologies (Spencer, 1985).

Initially, the international centers designed programs for new varieties and breeding lines of rice, maize, wheat, potatoes, and so on, which were tested in a number of countries. In Asia, this approach was successful with wheat and rice. However, the small-scale units and diverse farming systems frustrated this approach in Africa. There was a need to take the research itself (rather than its output) closer to the targeted environment (Jha, 1987). There are, however, fewer potential resources for generating new technology for the smaller-scale units. Experience has also shown that it is more difficult for research products to be transferred and adapted to such units.

One context in which the issues of diversity, indigenous technical knowledge, and links to world science are clearly intertwined at the international level is the case of small, low-income developing countries. For example, Guinea-Bissau, which has fewer than a million inhabitants, has over twenty ethnic groups, all speaking different languages. Tiny São Tomé e Príncipe has agroecological zones that range from dry savannah to rainforest to temperate equatorial montaine. The scale of these two countries results in economic constraints and additional costs for both the internal generation of technologies and the transfer of technology.

Small countries are less likely to have extensive, homogeneous, highly productive agricultural zones. The totality of their production in large-scale commodities does not provide any comparative advantage in relation to the larger-scale producers of these same commodities. The comparative advantage of small countries is precisely in exploiting their diversity, rapidly changing to exploit new niches within the world agricultural economy, and maximizing the use of local knowledge on diversity and new products specific to their respective environments. This comparative advantage must then be linked to the global agricultural research system to acquire the technology to improve the production of major staples and to increase the economic value of their distinctive crops.

At the macrolevel, the limited capacity of small countries to generate technology has also meant that they have been the least able to borrow and adapt technology. To borrow effectively, a small country needs to develop and institutionalize its own national research capacity, but how will those research organizations be structured? Where will the resources come from, and what level of research effort is required? How will they be oriented, and what strategies should they adopt?

The agreed-upon premises in discussions of agricultural research and development in small, low-income developing countries are that much of the new technology will need to be borrowed from external sources, and the countries themselves may not be able to support the research programs needed to encourage agricultural growth and manage the national resource base. By carefully considering where national agricultural research systems in small countries are concentrating their efforts and how they can make more effective use of knowledge sources outside the research system, it is hoped that strategies can be identified to allow small countries to sustain the scientific capacity essential for agricultural growth and conservation of resources.

Defining Small Countries and Their Research Systems

Since the issues of diversity, scale, and farmer knowledge are closely linked in the establishment of viable research systems and strategies for small, low-income countries, a more exact definition of the countries involved is needed. In examining the problems of organizing agricultural research in small countries, there have been consistent difficulties in defining the concepts of size and scale and the relevant units to which they are applied.

First, there has been a tendency to conflate small countries with small national agricultural research systems. This is confusing because most developing countries have small national research systems, and many of the organizational problems of research are those of small national systems in general. With fewer than two hundred national agricultural researchers, Zaire, for example, faces the problem of reorganizing a small research system into one with sufficient scope and capacity to develop this large country's rich agricultural potential. The problems of institutional development and organization of small national research systems apply to nearly all low-income developing countries, large and small.

The problem of agricultural research in small countries is defined not by the size of the national research system but by the size and resource base of the country, placing constraints on the development of a national system and structuring a set of choices for the organization and strategy of agricultural research. Our first step is to identify a set of low-income, small developing countries where agriculture is central to national development. Vernon Ruttan has suggested that research systems in fifty or so of the smallest low-income countries will be structurally and functionally different from the systems in larger countries. Ruttan compares these smaller systems to a branch station within the Netherlands or to the U.S. state of Minnesota (Ruttan, 1985).

Five criteria were selected as critical for defining the special conditions of agricultural research systems in small developing countries. The parameters were then applied to all independent countries, using 1980 statistics for the relevant areas of interest. The resulting group of fifty small countries was derived from a matrix of all countries with fewer than five million inhabitants (1980 census data) that fell within at least four of the five parameters. (All figures used in defining the parameters are for 1980 and in constant 1980 U.S. dollars. See Appendix 1.1.)

Three generalizations apply to the research systems of the fifty smallest countries:

- Small-country research systems have higher investment costs per hectare than large ones to achieve an equal level of effectiveness.
- Small countries with great agroclimatic variations will face higher costs in developing appropriate agrotechnologies than those with less variation.
- Small countries must organize research efforts to maximize potential to borrow technology and accept that much of their own technology-generating efforts will result in spillover benefits.

The costs and benefits of research in small countries, along with the very low level of available resources, make their choices for research allocation and the direction of their research effort more difficult. Because these restrictions are binding in terms of the absolute size and level of effort, the range of choice for these countries will be different as well. Hence, greater attention must be given to the links of the research system with world science and local knowledge (Gilbert and Sompo-Ceesay, 1988).

Thus far, the problem of agricultural research in small countries has been generally defined as one of borrowing and adapting technology from outside (Javier, 1988). The fundamental strategies for borrowing have been identified and exist in one form or another in many parts of the developing world. Technological borrowing can take the form of a regional network among small countries, a regional network (Plucknett and Smith, 1984) including larger countries as well as other small countries, commodity networks, networks based on political and financial links (as in francophone Africa), the international centers of the CG system, and links with a large neighbor with a comprehensive research system. Regional research institutions such as the Caribbean Agricultural Research and Development Institute (CARDI) based in Trinidad and Tobago (serving the CARICOM [Caribbean Community] region) execute regional programs and collaborative research with member national agricultural research systems.

The conditions that favor each type of network and borrowing strategy in relation to the type of commodity being borrowed and the research required to borrow and adapt it (as well as the edaphic and geographic conditions) remain to be evaluated and are a major component of the International Service for National Agricultural Research (ISNAR) study on this topic.

Indigenous Technical Knowledge

While borrowing agrotechnology will be the common strategy for research systems in small countries, a neglected aspect of research and a source of new agricultural expertise that can be placed at the service of rural development is farmer knowledge—particularly useful for acquiring basic information about the agricultural and natural resource base (Richards, 1985). Small countries as well as large ones can tap this rich source of locality-specific information about agriculture for identifying new or potential areas of research and sustainable systems for managing basic agricultural resources.

This consideration of the problems of developing national agricultural research systems and extending the impact of rural agricultural science and technology to small, resource-poor countries proposes a dual focus for national systems in small countries: (1) the need to link the borrowing of externally generated technology with in-depth systematization, and (2) the use of local farmer knowledge.

Defining Indigenous Technical Knowledge

Indigenous technical knowledge is vast. It includes most of the agricultural farming techniques in use (Tripp, 1988:5), yet there are clear boundaries between farming knowledge and agricultural science, which, when properly understood, serve as the basis for the complementary focus necessary for extending agricultural research to small countries. The two systems of knowledge are defined by their respective goals and scope. Indigenous technical knowledge is a body of information applied to the management of natural resources and labor within very specific plots. Farmers' knowledge about the specific conditions in which they produce may be more exact than the knowledge of trained researchers who are producing new crop varieties or other technologies for these environments. This is not a failure of the research system or the idealization of the low-resource farmer, but a recognition of the division of labor between scientific agricultural research and the empirical knowledge that farmers acquire in order to produce with available resources.

Scientific research in agriculture produces new knowledge that does not result directly from the act of farming. The validity and utility of scientific information about farming are measured by its ability to produce similar results under similar conditions throughout the world. In the past, the international centers have placed strong emphasis on the external inputs of fertilizers, weed control, and irrigation which provide greater control over the farm environment. The success of the CG centers has been notable in areas where large-scale farming conditions have been similar or where it is feasible to use external inputs to produce the necessary environment for increased production. The international agricultural research centers' new focus on smaller-scale units and more specific and difficult agroecologies has increased costs and casts doubts on the feasibility of the earlier strategy.

To reach the bulk of the rural poor, we need a new strategy to link the generality of scientific agricultural research with the highly specific and empirical knowledge of local conditions. Linking farmer knowledge with world science requires a better understanding of both the role of scientific agricultural research and the limits of empirical locality-specific farmer knowledge. It is not possible for research within international centers or even within many national research programs to produce a technical package or set of recommendations for each farming system. In this sense, there is the danger that farming systems research, when seen as a way of improving research methodologies and identifying key issues that agricultural research systems must face, has extended the scope of agricultural research into areas where it cannot produce results (Tripp, 1986).

Some systems of farmer knowledge have already been incorporated into farming systems research to identify areas where technological inputs are needed to improve the productivity or sustainability of agriculture in a given area. The implicit expectation that a national system can target its research to specific farming systems places an undue burden on the national system in an area in which formal research programs have no comparative advantage. The production of generalized new information is the appropriate task of research, while the testing of its usefulness within specific agroecologies and farming systems is the task of the farmer.

Incorporating Farmer Knowledge into the Agricultural Research Systems of Small Developing Countries

Incorporating farmer knowledge into agricultural research depends in part on identifying areas of complementarity between the two knowledge systems. For this reason, the scope of research in small-country national agricultural research systems has been a key first step in the research. Following analysis of the existing scope of national research in fifty small countries, it became clear that there were distinct gradients and flows in agrotechnology and information between national research systems. These were dependent to a large extent on the type of technology and commodity. Based on the survey, they have been grouped into the following categories: global staples, traditional exports, minor food crops, high-input, nontraditional exports, natural resources management, livestock research, socioeconomics, and rural engineering.

Global staples: This category includes major food crops with a global distribution in terms of both production and sources and transfer of new technology. Typically, these crops are the focus of work by the international agricultural research centers. National systems and the private sector are also significant sources of technology information on these commodities.

Traditional exports: Included in this category are fiber crops, gum crops, oil crops, stimulants, medicinal plants, and spices. These crops are historically produced for the global market, and research is distributed worldwide with important contributions from the private sector.

Minor food crops: This category comprises crops that are locally important to the food-producing sector within a country and are not a major component of a country's agricultural exports. New technology on these crops is less readily available or is not specifically targeted to developing countries.

High-input, nontraditional exports: Crops grown primarily for export to consumers in developed countries are included in this category. Major emphases in production are quality, uniformity, and timing, which necessitates a high level of inputs, controlled conditions, and special handling. Postharvest considerations are particularly important. The private sector plays a major role in the generation and transfer of technology for these crops. Small-scale production may be economical provided the distribution and marketing system is present.

The private sector is a major source of new technology for this group of commodities.

Natural resources management: Included in this category are research topics that are not commodity based and are concerned with managing an existing resource base such as soils, water, plants, or fish stocks to increase, extend, or conserve the productivity of the particular resource. There is an inherent logic in conducting this type of research within the country; it can be complex, however, even at an adaptive level. Nongovernmental organizations (NGOs) have played an important role in this type of research, with recent involvements of the international centers.

Livestock research: This category comprises all topics related to animal production: zootechnology, animal diseases, fodder, nutrition, and livestock management. Principal sources of technology are international centers and veterinary services in more developed countries.

Socioeconomics and rural engineering: This category includes research topics dealing with the management and resource allocation of farm enterprises. It covers socioeconomic studies of production; farmers' choices, preferences, and constraints; farming systems research; marketing research; storage; and farm structures. This research is country-specific, employing more widely applicable methodologies.

Table 1.1 indicates the twenty-nine countries for which we were able to obtain comprehensive data on the scope of national agricultural research. Table 1.2 indicates the areas in which national programs are conducting research. It shows that small research systems have more programs concentrated in the global staples for which there is a well-structured system for developing and transferring technologies, training, and funding through the international centers, the donors, and in some cases the private sector. A similar concentration can be seen in traditional exports, where a well-established system of research and development exists in nationally based public and private research organizations. The flow of new technologies is rapid between centers, and the private sector is involved in both research and transfer.

Initial findings from a global study under way to assess the intensity of small-country agricultural research system efforts in these areas (Eyzaguirre, 1990) indicate that a proportionately smaller share of national research efforts is allocated to topics in natural resources management and socioeconomics and postharvest research. In-depth

Table 1.1. Summary report: Countries undertaking research under the different categories of research areas/topics identified for the study of national agricultural research systems in small countries, 1985–89

Global staples	Traditional exports	Minor food crops	High-input, non-traditional exports	Livestock	Socioeconomics and rural engineering	Natural resource management
Tonga	Seychelles	Fiji	Jamaica	Swaziland	Gambia	Tonga
Swaziland	Mauritius	Somalia	Swaziland	Somalia	Swaziland	Swaziland
Suriname	Tonga	Swaziland	Fiji	Nicaragua	Lesotho	Solomon Islands
Somalia	Solomon Islands	Jamaica	Lesotho	Seychelles	Sierra Leone	Seychelles
Solomon Islands	Fiji	Seychelles	Botswana	Rwanda	Papua NG	Rwanda
Sierra Leone	Guyana	Rwanda	Tonga	Fiji	Botswana	Paraguay
Seychelles	Gambia	Bhutan	Bhutan	Mongolia	Mongolia	Papua NG
Rwanda	Kiribati	Tonga	Paraguay	Jamaica	Congo	Panama
Paraguay	Papua NG	Guyana	Gambia	Mauritania	Mauritania	Mongolia
Papua NG	W. Samoa	Panama	Papua NG	Lesotho	Guyana	Mauritania
Panama	Nicaragua	Sierra Leone	Guyana	Panama		Liberia
Nicaragua	Liberia	Mauritania	Somalia	Papua NG		Lesotho
Mauritania	Rwanda	Lesotho	Panama	Gambia		Laos, PDR
Liberia	Jamaica	Papua NG	Rwanda	Congo		Kiribati
Lesotho	Burundi	Nicaragua	Cape Verde	Cape Verde		Jamaica
Laos, PDR	Cape Verde	Paraguay		Burundi		Guyana
Jamaica	Swaziland	Gambia		Botswana		Gambia
Guyana		Solomon Islands				Fiji
Gambia		Congo				Cape Verde
Fiji		Cape Verde				Botswana
Congo		Burundi				
Cape Verde		Botswana				
Burundi		Kiribati				
Botswana						
Bhutan						
Total 25	17	23	15	17	10	20

Table 1.2. Categories of research domains

Global staples	Traditional export crops	Minor food crops	High-input, nontraditional export crops	Livestock	Socioeconomics and rural engineering	Natural resources management
Beans	Bananas	Apples	Asparagus	*Small ruminants*	Agricultural wastes	*Fisheries*
Cassava	Cashew nuts	Barley	Broccoli	Goats	Agroindustries	*Forestry*
Cowpea	Cinnamon	Breadfruit	Brussels sprouts	Sheep	Agroprocessing	Agroforestry
Groundnut	Cloves	Broad and mung beans	Cardamom	*Large animals*	Farming systems research	Genetic resources
Maize	Cocoa	Cabbage	Citrus (limes, grape-fruit)	Camels	Farm management	Plant pest and disease management
Potatoes	Coconuts	Carrots	Flowers/ornamentals	Cattle	Farm structures	*Land use and water management*
Pulses	Coffee	Castor beans	Fruits	Donkeys	Irrigation	Soil (fertility, erosion, conservation)
Rice	Cotton	Data palms	Ginger	Horses	Machinery and tools	Range and pasture
Sorghum	Oil palm	Figs	Grapes	*Small stock*	Marketing research	Water resources management
Soya	Rubber	Fruits (local use)	High-value vegetables	Chickens	Postharvest and storage	
Wheat	Sisal	Garlic	Jojoba	Ducks	Rural engineering	
	Sugar	Lentils	Kava	Swine		
	Tea	Millet (*Eleusine, Digitaria*)	Litchi	Turkeys		
	Tobacco	Mustard (seed)	Mangoes	*Animal health*		
		Oats	Melons	*Feeds and nutrition*		
		Okra	Palm hearts	*Animal breeding*		
		Onions	Papaya	*Wildlife management*		
		Pandanus	Passion-fruit	*Aquaculture*		
		Pears	Peaches			
		Peas (garden)	Pineapples			
		Peppers	Plums			
		Pigeon peas	Pyrethrum			
		Plantain	Quinquina			
		Radishes	Ramie (textile fiber)			
			Sour sop			

studies under way in seven countries (Honduras, Jamaica, Togo, Sierra Leone, Lesotho, Mauritius, and Fiji) show that research in natural resources management does not extend much beyond analyses of soil fertility and laboratory services. In several countries, such as Lesotho, Jamaica, and Honduras, conservation of natural resources and diversification of the agricultural resource base are major policy objectives. The current flows of new information about these crucial research areas also indicate that new knowledge has to be tested and adapted to local conditions. A thorough understanding of the existing resource base and farming systems is the crucial first step.

The hypothesis being tested in several of the case-study countries is that socioeconomics, rural engineering, and natural resources management are likely to be areas where national systems can increase the benefits from borrowing technologies and contribute new knowledge on the management of agricultural resources in small-scale production systems.

In Sierra Leone, the national research programs have made significant and highly successful efforts at integrating a farming systems perspective into their programs (Dahniya, 1990), and farmer knowledge has been a key input. It is still too early to tell how this contribution will affect research. Perhaps it may be in the area of documenting the resources that already exist and the ways farmers can economically cope with resource constraints. Paul Richards found that recent farming systems research in West Africa seems to have merely reinvented traditional agriculture; yet this recovery of existing systems of production may be the basis for future technological progress (Richards, 1983).

It is also clear, based on work conducted in Honduras, that farmer knowledge systems are themselves incomplete and biased. Farmers tend to know a great deal about local agroecosystems. Their knowledge of plants, growth stages, and plant interaction is impressive, but they know less about insects and plant pathology, for example. The Honduras case illustrates that gaps in local farmer knowledge that are not compensated for in formal research and extension can lead to a widespread misuse of pesticides (Bentley, 1989).

In the categorization of research programs in the sample of small-country national research systems, certain categories clearly allow for greater input by local farmers—for example, natural resources management, socioeconomics, postharvest and rural engineering, and minor food crops. These are also areas in which technology and information are not as readily available or as easily transferable to the national systems.

Developing effective technologies with farmer input often requires that the work be done locally, in many cases under farm conditions. Increased attention to those areas to which farmers can contribute may not only be necessary, it may also lead to increased benefits from borrowing the appropriate technology where it is readily available. New opportunities for high-value crops that are economical when produced at small scales will require extensive national research in the areas of socioeconomics, marketing, and postharvest technologies, even if much of the production technology already exists. A major hypothesis that emerges from the global study of national agricultural research systems in small countries where low-income farmers predominate is that the areas where the national systems should dedicate increased efforts are also areas where they may have a comparative advantage through integrating local farmer knowledge.

This focus on small countries moves the work of agricultural research closer to the smaller-scale units of production with fewer external inputs, where diversity in agroecological conditions and farming practices is the norm and the desired end. It allows us to consider the relationship between agricultural science and indigenous farmer knowledge and their complementarity in the process of technological innovation; and it has the potential to bring us closer to an understanding of the new division of labor within the global research system.

Conclusions

The first phase of work by the global agricultural research system—maximizing production of the major grains and staples in the regions with the greatest potential to increase yields through external inputs—has been successful. Future work on biotechnology and genetic engineering may rely a good deal more on indigenous knowledge of little-known cultivars, varieties, and species.

The emerging agenda of sustainable production—protecting and successfully managing agricultural and environmental diversity and reaching the rural poor in smaller-scale social units and more variable agroecologies—is in fact the fundamental agenda of national research systems in developing countries. But they will not be able to address these questions without strong links to the global agricultural research system. The focus on the organization and strategies of agricultural research systems in small countries may make a significant contribution to the work of the international centers by providing a clear context in which to address the issues that were avoided in the earlier phase because of their complexity and cost.

The initial findings of the national systems study indicate that a shift of emphasis—increasing the efforts in the socioeconomics, rural engineering, and natural resources management categories—will enable national systems in small countries to do two things. First, these systems will be able to borrow technology more efficiently through a better knowledge of the needs and ability of farmers to use it economically. Second, socioeconomics, rural engineering, and natural resources management are research categories in which farmer knowledge can contribute much to the development of new technologies. This has been demonstrated in the work of the International Potato Center in postharvest technologies and in the identification of traditional crops that have potential to become high-value, nontraditional exports.

In linking the work of agricultural science with farmers' knowledge of production, farm management, and the environment, a small country does not suffer any limitations of scale. National scientists in some of the smaller research systems, such as that of Sierra Leone, have made significant breakthroughs based on the synthesis and development of traditional rice production systems that maximize diversity in both cultivars and production techniques. In many of the countries in our sample, agricultural development policies include diversification of production and exports as a key objective. Identifying new crops that can diversify the production base and perhaps claim a niche as high-value exports will benefit from tapping farmers' fund of knowledge about plants, cultivars, and their uses, as well as the value they have traditionally placed on diversified production.

Appendix 1.1 Criteria for Selecting Small Countries and Their Research Systems

A. Population five million or less (1980). Countries with less than five million inhabitants pose a special case for the structuring and organization of research systems (Ruttan, 1986:321). Major international organizations that compile statistical data on small developing countries have consistently used five million as the cutoff point in defining small countries where size poses a distinct problem for economic development. (Commonwealth Secretariat, 1988, basic statistical data on selected countries with populations of less than five million.)

B. *Per capita income is less than $2000 (1980 U.S.$).* This is used to define low-income countries for our purposes and to direct attention to those countries most in need of assistance.

C. *Economically active agricultural population is greater than or equal to 20 percent of the total economically active population.* This parameter selects countries with relatively large agricultural populations. This means that a significant sector of the population will be affected by improvements in agricultural productivity and agricultural income.

D. *Agricultural GDP per economically active agricultural population is less than or equal to $2000 (1980 U.S.$).* This parameter selects countries whose agricultural sector is made up largely of low-income farmers. This responds to the CGIAR's priority for assisting agricultural sectors based on need and equity.

E. *Agricultural domestic product is greater than or equal to 20 percent of the GDP.* This parameter ensures that selected countries have significant agricultural sectors. Promoting agricultural productivity through research is therefore likely to have an important impact on national development. It also eliminates small but mineral-rich countries that have the financial resources to invest in agricultural research and training should agriculture assume importance in their development plans.

Note: This section, defining and selecting a sample group of small countries, was the result of work by B. Folger of ISNAR's research staff and will be further developed into a data base on small-country national agricultural research systems.

Appendix 1.2. Statistical parameters for selecting sample of small countries (1986 indicators, FAO)

Country	Population (number × 1000)	Per capita income (U.S.$)	Agricultural contribution to GDP (%)	Agricultural per capita income	Percentage of population in agriculture
Belize	167	1228	21	1901	36
Benin	4177	316	49	500	65
Bhutan	1389	151	51	189	91
Botswana	1117	877	5	178	66
Burundi	4864	254	51	269	92
Cape Verde	338	334	25	482	46
Central African Republic	2638	342	41	428	67
Chad	5140	163	46	270	78
Comoros	458	332	42	378	81
Congo	1787	1118	6	261	61
Djibouti	360	956	4	105	79
Dominca	77	1325	19	1762	31
El Salvador	4846	821	20	1362	39
Equatorial Guinea	401	204	46	370	60
Fiji	703	1879	17	2299	42
Gambia	767	237	15	94	82
Grenada	97	1330			
Guinea-Bissau	908	184	45	215	80
Guyana	971	535	21	1329	24
Honduras	4532	798	27	1243	57
Jamaica	2371	1026	6	462	29
Kiribati	65	323	27	1418	17
Laos, PDR	3683	767	65	1402	73
Lesotho	1583	181	17	78	83
Liberia	2249	482	34	597	72
Maldives	189	460	13	251	66
Mauritania	1814	465	30	652	66
Mauritius	1051	1328	13	1796	25
Mongolia	1964				34
Namibia	1650	560			38
Nauru	8				
Nicaragua	3384	858	23	1557	42
Panama	2227	2300	9	2216	28
Papua New Guinea	3605	704	34	708	71
Paraguay	3807	909	29	1623	47
Rwanda	6312	293	40	255	92
São Tome e Príncipe	100	440	27	457	67
Seychelles	60	2962	7		
Sierra Leone	3755	303	42	553	65
Solomon Islands	279	427			48
Somalia	6623	147	65	319	73
St. Lucia	130	1451	12	1258	30
St. Vincent	105	1118	15	1213	31
Suriname	380	2589	9	4441	18
Swaziland	688	613	23	495	70
Togo	3055	322	32	351	71
Tonga	112		32	2	17
Tuvalu	8				
Vanuatu	147	469	34	1066	48
Western Samoa	165	655	31	4185	15

Source: ISNAR small countries data base.

Bibliography

Anderson, Jock R., George Antony, and Jeffrey S. Davis. "Research Priority Setting in a Small Developing Country: The Case of Papua New Guinea." Paper presented at ISNAR/Rutgers Agricultural Technology Management Workshop, New Brunswick, N.J., July 1988.

Bentley, W. J. "What Farmers Don't Know Can't Help Them: The Strengths and Weaknesses of Indigenous Technical Knowledge in Honduras." *Agriculture and Human Values*, 6(3):25–31, 1989.

Chambers, Robert, and Janice Jiggins. "Agricultural Research for Resource Poor Farmers: A Parsimonious Paradigm." Discussion Paper, IDS, 1986.

Collinson, M. P. "Farming Systems Research: Procedures for Technology Development." *Experimental Agriculture*, 23:365–86, 1987.

Crusol, Jean, and Louis Crusol. "A Programme for Agriculture in Island Plantation Economies." University of the French West Indies, Martinique. *World Development*, 8:1027–33, 1980.

Dahniya, M. T. "Sierra Leone's Agricultural Research System: Science and the Farmer, a Small-Scale Approach to Technological Development in Agriculture." ISNAR, Special Project on Agricultural Research in Small Countries, 1990.

Davis, Carlton G. "Agricultural Research and Agricultural Development in Small Plantation Economies: The Case of the West Indies." *Social and Economic Studies*, 24(1):117–52, 1975.

de Groot, Peter. "New Life in Old Crops." *New Scientist*, 118(1607), 1988.

Echeverria, Ruben G. "Agricultural Research Policy, the Small Country Case." Unpublished report, Department of Agricultural and Applied Economics, University of Minnesota, 1986.

Elliott, Howard. "Applying ATMS Approaches in Widely Different Systems: Lessons from ISNAR Experience." Paper presented at ISNAR/Rutgers Agricultural Technology Management Workshop, New Brunswick, N.J., 1988.

———. "Agricultural Research and the Policy Environment." Paper presented at the International Agricultural Research Management Workshop, ISNAR, 1990.

Ellman, Antony. "Technology Transfer in Smallholder Farming Systems: A Programme of Promotion in Africa and the South Pacific." *Agricultural Administration and Extension*, 25:143–60, 1987.

Eyzaguirre, P. B. "The Ecology of Swidden Agriculture and Agrarian History in São Tomé." *Cahiers d'études africaines* (Paris), 101–2, 1987.

———. "Competing Systems of Land Tenure in an African Plantation Society." In R. Downs and S. Reyna, eds., *Land and Society in Contemporary Africa*. Durham, N.H.: University Press of New England, 1988.

———. *Developing Appropriate Strategies for Agricultural Research in Small Countries*. The Hague: International Service for National Agricultural Research, 1990.

Fresco, L. O., and E. Westphal. "A Hierarchical Classification of Farm Systems." *Experimental Agriculture,* 24, 1988.

Gilbert, Elon H., and M. S. Sompo-Ceesay. "Dealing with the Size Constraint: Strategies for Technology Management in Small Agricultural Research Systems." Paper presented at Rutgers/ISNAR Agricultural Technology Management Workshop, New Brunswick, N.J., 1988.

Harwood, R. *Small Farm Development.* Boulder, Colo.: Westview Press, 1979.

Horton, Douglas E. "Assessing the Impact of International Research: Concepts and Challenges." Paper presented at Rutgers/ISNAR Agricultural Technology Management Workshop, New Brunswick, N.J., 1988.

ISNAR. *Agricultural Research Policy and Organization in Small Countries.* Workshop proceedings, Wageningen, September 1984. The Hague: ISNAR, 1985.

Jahnke, H. E., D. Kirschke, and J. Lagemann. "The Impact of Agricultural Research in Tropical Africa." CGIAR Study Paper no. 21. Washington, D.C.: World Bank, 1987.

Javier, E. Q. "The Small Country Problem: A Reflection." ISNAR Discussion Paper. The Hague: ISNAR, 1988.

Jha, Dayanatha. "Strengthening Agricultural Research in Africa: Some Neglected Issues." *Quarterly Journal of International Agriculture,* 26(3), 1987.

Merrill-Sands, D. *The Technology Applications Gap.* Rome: FAO, 1986.

Moock, J. L., ed. *Understanding Africa's Rural Households and Farming Systems.* Boulder, Colo.: Westview Press, 1986.

Norman, D. W., E. B. Simmons, and H. M. Hays. *Farming Systems of the Nigerian Savanna.* Boulder, Colo.: Westview Press, 1982.

Payne, Anthony J. *The Politics of the Caribbean Community 1961–1979; Regional Integration amongst New States.* Manchester: Manchester University Press, 1980.

Persaud, Bishnodat. "Agricultural Problems of Small States, with Special Reference to Commonwealth Caribbean Countries." *Agricultural Administration and Extension,* 29, 1988.

Pitt, David. "Commission of Ecology: Working Group on Islands." International Union for the Conservation of Nature and Natural Resources. Gland, Switzerland, 1986.

——. "Small Island Economies Face Bigger Problems." *Ceres* 124 (vol. 21, no. 4), 1988.

Plucknett, D. L., and N. J. H. Smith. "Networking in International Agricultural Research." *Science,* 225(4666), 1984.

Rhoades, Robert E., and Robert H. Booth. "Farmer-Back-to-Farmer: A Model for Generating Acceptable Agricultural Technology." *Agricultural Administration,* 11(2):127–37, 1982.

——. "Farmers and Experimentation." Agricultural Administration (Research and Extension) Network Discussion Paper 21. London: Overseas Development Institute, 1984.

Richards, Paul. "Farming Systems and Agrarian Change in West Africa." *Progress in Human Geography,* 1:1–39, 1983.

——. *Indigenous Agricultural Revolution.* London: Hutchinson, 1985.

Ruttan, Vernon W. *Small Country Agricultural Research System: Agricultural Research Policy and Organization in Small Countries.* The Hague: ISNAR, 1985.

——. "Toward a Global Agricultural Research System: A Personal View." *Research Policy, 15:* 307–27, 1986.

——. "Toward a Global Agricultural Research System." In *The Changing Dynamics of Global Agriculture: Research Policy Implications for National Agricultural Research Systems.* The Hague: International Service for National Agricultural Research, 1988.

Schuh, G. E. "Sustainability, Marginal Areas, and Agricultural Research." Paper prepared for International Fund for Agricultural Development. Rome, 1988.

Selwyn, Percy. *Development Policy in Small Countries.* IDS. London: Croom Helm, 1975.

Spencer, D. S. C. *Agricultural Research in Sub-Saharan Africa: Using the Lessons of the Past to Develop a Strategy for the Future.* Niamey: Committee on African Development Strategies, 1985.

Technical Advisory Committee of the Consultative Group on International Agricultural Research (TAC). "Sustainable Agricultural Production: Implications for International Agricultural Research." (AGR/TAC:IAR/87/22 Rev. 2.) Technical Advisory Committee, Rome, 1988.

Tripp, R. "Farmer Participation in Agricultural Research: New Directions or Old Problems?" ISNAR Staff Seminar, 1988.

Turner, B. L., and S. B. Brush, eds. *Comparative Farming Systems.* New York: Guilford Press, 1987.

Ward, R. G., and A. Procter, eds. *South Pacific Agriculture: Choices and Constraints.* South Pacific Agricultural Survey 1979. Canberra: Asian Development Bank and ANU, 1980.

Zuleta C., Juan Carlos. "Country Size, Level of Development, Relative Importance of Agriculture and Agricultural Research in LDCs." Staff Paper, Department of Agricultural and Applied Economics, University of Minnesota, September 1986.

2

Conserving and Increasing On-Farm Genetic Diversity: Farmer Management of Varietal Bean Mixtures in Central Africa

Joachim Voss

Introduction

This chapter assesses the feasibility of developing technologies that enhance, rather than reduce, the on-farm genetic diversity of bean varietal mixtures in central Africa. My research in this region demonstrates that, for reasons of food security and yield, farmers strongly prefer to grow mixtures, and that their seed selection procedures maintain a higher degree of genetic diversity than would be the case under natural selection. The results of on-station research show that mixtures effectively reduce disease severity and spread and may have further yield-enhancing effects through the association of beans of different plant architectures, however, researchers' and market preferences for pure lines limit the utilization of mixtures.

The potential of this farming practice has not gone unnoticed. As early as 1976, Neeley et al. recommended to the Agricultural Development Council that breeding programs should more fully exploit the traditional farmer strategy of increasing genetic diversity by growing many varieties in the same field: "The major efforts in most breeding programs have been directed toward the development of a few 'elite' high-yielding varieties. This practice often leads to the release of genetically narrow based material which leaves crops at the mercy of abnormal fluctuations in climatic conditions and infestation from new pests or pest biotypes. Alternative breeding philosophies and programs do exist and should be adopted more widely in national and international agricultural research institutions" (1976:2). In the last few years, concern for the conservation of genetic diversity has increased tremendously. Most of this concern has had a laudable focus on the creation of nature preserves and germ plasm collection and storage (Abramovitz, 1989; McNeeley et al., 1990; McPherson, 1985).

Relatively little attention, however, has been paid to the possibility of devising research strategies that would both increase the quantity and improve the quality of the genetic diversity being conserved by farmers. The current tendency to collect so-called land races for gene banks and replace them with elite lines has had the further unintended consequence that farmers' knowledge of the characteristics and local adaptation of cultivars is becoming narrower. Stressing that almost all the African economies are highly dependent on agriculture, Juma emphasized the crucial importance of biological diversity for the economic renewal of Africa. He claimed that breeding programs have led toward a narrowing of the genetic base in Kenya (1989:21). He contrasted this tendency with the prevailing preference for diversity on the part of local producers and consumers, citing the specific case of beans (ibid.:22).

Historically, in most of highland Africa beans were grown as varietal mixtures by women, usually in association with other crops. In countries like Kenya and Uganda, where colonial agricultural policy was powerful and influential, this practice was systematically discouraged—either because of a desire to promote the production of uniform "export-quality" beans or because the cultivation of pure "improved" lines was considered to be more modern and ultimately more productive (C. Robertson, personal communication, 1988). Consequently, mixtures today are commonly grown only for subsistence and in the more remote regions of these countries.

In the Great Lakes region of central Africa (Rwanda, Burundi, and the Kivu Provinces of Zaire), the growing of mixtures remains nearly universal. Only in North Kivu, where most of the beans for export to the Kinshasa market are produced, are beans of uniform color grown. Upon closer examination, even these apparently pure lines turn out to be less complex mixtures of three to five varieties of similar color.

Recognizing the importance of at least maintaining and preferably broadening genetic diversity, the bean program of the Centro Internacional de Agricultura Tropical (CIAT; the international agricultural research center with a global responsibility for bean germ plasm and research) is trying to ensure that research and development efforts do not lead to the kind of genetic erosion that has taken place in Kenya, Uganda, and, even more spectacularly, Latin America, the center of origin of the species *Phaseolus vulgaris*.[1] The practice of growing varietal mixtures in the Great Lakes region represents an opportunity to increase the stability and productivity of these mixtures, which can be accomplished through the addition of new varieties that incorporate desirable characteristics of disease resistance, higher yield potential, and farmer and consumer acceptability. With mixtures, breeders are spared the difficult task of trying to incorporate all resistances and potentials into a single variety and can instead stack up such characteristics by incorporating multiple new lines into the mixtures.

When CIAT first began its work in Africa, it had virtually no experience with farmers growing varietal mixtures. Therefore, one of the first tasks was to document farmers' practices[2] in relation to bean mixtures and to analyze the implications of these practices for the research strategy that was to be developed for the Great Lakes region.

Starting in 1983, an interdisciplinary team of five scientists—comprising a breeder, an anthropologist, a plant pathologist, a nutritionist, and an agronomist—was placed in the Great Lakes region. The team's major objective has been to develop agricultural systems and technologies that will increase the productivity and stability of common beans *(Phaseolus vulgaris)* for small farmers in the region. To accomplish this, CIAT works with national programs and development projects on methodology, research, and extension strategy improvement (CIAT, 1985:274).

The Role of On-Farm Research

On-farm research is a key component in realizing the project's objectives in an efficient and equitable manner and also in keeping the research relevant. The collection of diagnostic information from farmers, their evaluation of new technologies, and their participation in

[1] For an excellent summary discussion of the issue of plant genetic diversity and sources of an access to genetic resources, see Ferguson and Sprecher, 1987.

[2] References to farmers throughout this paper refer primarily to women, since most of the food-crop production falls on their shoulders. Men, to a variable and lesser extent, are also involved in food-crop production, particularly in field preparation, in decisions on what to plant where, and sometimes as a source of new varieties they have come across when traveling.

research all play important roles in the project. It is at this level that many of the project's research priorities are defined and the technologies and approaches generated by the project are ultimately tested.

To a large extent, the production and consumption problems that farmers have identified as most pressing have determined research priorities. Western agriculture is highly specialized and market oriented. As a consequence, production is clearly separated from consumption patterns and production itself involves a very limited number of crops and activities. However, most production by small African farmers is consumption driven. In Rwanda, for example, over 85 percent of production is consumed by the farm household. Thus, consumption needs and preferences must be understood in order to comprehend the rationality behind production. Staggered planting dates and the types of crops grown in association are determined by consumption needs as well as by the need to minimize risks.

Because of the multifaceted nature of farmers' production and consumption patterns, several approaches are required to analyze these complex systems adequately, providing a focal point for interdisciplinary discussion. For example, the anthropologist may determine that farmers consider bean disease problems during periods of high rainfall a major production constraint. The pathologist might then identify the disease problem as a combination of aschochyta, anthracnose, and angular leaf spot. Research may show that the severity of these diseases can be reduced by planting after the heaviest rains have passed; however, the nutritionist may show that this is not a viable option because beans are one of the first foods to become available to relieve the food scarcity following the dry season. Thus, other options, such as varietal resistance or seed treatment, need to be considered in the same interdisciplinary manner.

Beans in the Great Lakes Region

The Great Lakes region is at the heart of the Central African Highlands, on either side of one branch of the Rift Valley system. Running from north to south, the valley contains Lakes Edward, Kivu, and Tanganyika. Composed of high plateaus, volcanoes, and high mountain ridges on either side of the rift, it descends into savannah plains toward the east. The altitude ranges between 900 and 4500 m above sea level, and rainfall ranges from less than 1000 mm in the east and along the valley bottom to more than 1800 mm along the Nile-Zaire crest and in the area of the volcanoes. Beans are grown in all parts of the region at altitudes between 900 and 2400 m. The Central Plateau region of Rwanda and Burundi receives between 1000 and 1400 mm of rain (Sirven et al., 1974:25).

The two major cropping and rainy seasons are from mid-September to early January and from late February to early June; however, the intensity and duration of the rainy seasons vary considerably from year to year. The dry seasons are longer and more pronounced in the east. Given this tremendous variability in altitude, climate, and soils, many bean production problems and solutions are highly location-specific. It is difficult, if not impossible, to develop a single variety or technological intervention that will have a near-universal impact.

Much of the Great Lakes region supports Africa's highest population density—over 350 people per square kilometer of agricultural land, with a projected density of over 500 by the end of the century. Parts of the Kivu Provinces of Zaire are an exception, with a considerably lower population density and more extensive forms of agriculture. Physically, the Kivu Provinces are the best endowed of the region, with large areas of fertile volcanic soil. The main problems are a lack of well-developed infrastructure and a highly unequal land distribution. North Kivu in particular acts as an in-migration area for some of the excess Rwandan population.

Over 95 percent of the population is rural, with an average farm size of less than 1 ha in Rwanda and Burundi (Gahamanyi, 1985:4). In the most densely populated areas such as the Central Plateau and the shores of Lake Kivu, over 50 percent of the farms are smaller than 0.5 ha. The eastern part of the region is lower and hotter, with more intense dry seasons and generally larger farms averaging about 3.5 ha. The Central Plateau is characterized by rolling hills separated by marshes which provide a dry-season crop. Soil composition and fertility are extremely variable (Sirven et al., 1974:41).

In terms of land area cultivated, bananas are the dominant crop, followed by beans, sweet potatoes, cassava, and sorghum. The soils of the highlands of the Nile-Zaire crest have high organic content but are also high in acidity and aluminum. Bananas and beans predominate in the more fertile valleys; cassava and sweet potatoes are grown on the heavily eroded slopes; and maize, peas, beans, sorghum, wheat, and potatoes predominate in the higher areas. Rainfall is more intense than in most other regions, and lodging and hail damage are serious problems at certain times of the year. The western slopes down to Lakes Kivu and Tanganyika have rainfall similar to that of the Central Plateau. The major crops there are maize, beans, cassava, and bananas (Jones and Egli, 1984:26–32).

All the major types of beans are grown in the region: bush, semi-climbing, and climbing; however, climbing beans are grown mainly in high-rainfall areas and are little known elsewhere. Beans are typically grown as varietal mixtures and intercropped with a wide range of other crops, especially bananas, maize, sweet potatoes, peas, cassava, cocoyams, and, at higher altitudes, potatoes. North Kivu is again an exception, with considerable cultivation of almost pure lines—particularly mixtures of whites, yellows, and red types—for sale, and a much lower level of intercropping.

Owing to heavy population pressure and a scarcity of fertile land in Burundi and Rwanda, fallow periods have declined and bean production has expanded into marginal land, causing average yields to drop from 0.9 metric tons/ha to 0.7 metric tons/ha, while total output has barely kept up with a population increase of 3.5 percent (CIAT, 1984:274). Beans are the region's single most important source of protein, contributing some 45 percent of protein needs. They also provide approximately 25 percent of caloric requirements (CIAT, 1984:279).

Given that sparsely occupied land available for new settlement has now been exhausted, further increases in food production will have to be achieved through intensified production on existing farmland. Such intensification provides a major challenge because the reduction of fallow time presumably accelerates the decline in soil fertility if farming systems are not adjusted to fit this new reality.

On-Station Research on Bean Mixtures

National agricultural centers in Burundi, Malawi, Rwanda, Tanzania, and Zaire have carried out research on bean mixtures since the 1950s. The bulk of this research has been of two kinds: studies of effects of mixtures on severity and spread of common bean diseases, and population studies on the evolution of single varieties through several generations of competition or complementarity from other varieties in the mixture (CIAT, 1987a).

Research on effects of varietal mixtures on diseases clearly shows their beneficial effect in improving and stabilizing yields, partly by decreasing the severity and spread of diseases. Working in Tanzania, Lyimo and Teri (1984) found that both rust and angular leaf spot increased more rapidly in pure lines than in mixtures. The mixture yielded 13.8 percent over the mean yield of the mixture components in pure stand. The local mixture also had a yield advantage of 10.6 percent. Ishabairu and Teri (1984) found a yield advantage of 24 percent with a lower disease severity in the mixtures and concluded that "for developing countries, the use of cultivar mixtures is more advisable than the use of multilines" (p. 17).

Edje and Adams (1985), working in Tanzania, DeVos et al. (1983) in Burundi, and Trutmann and Mukeshi (1988) in Zaire reported similar results. Thus one can see that the early colonial attitude, presupposing the superiority of the Western practice of growing pure lines, has not been borne out.

Research on the evolution of component varieties within mixtures shows that one or a few varieties quickly dominate the mixture after a few generations. Work on mixtures carried out at CIAT headquarters (CIAT, 1987) showed that leaf area index and seed size were the main determinants in a variety's competitiveness in mixtures. Varieties with high leaf area index and small seeds tended to predominate. Under disease pressure, susceptible varieties did much better in mixtures than in pure stands. One rust-susceptible variety, BAT 1297, had a mean yield gain of 100 percent in mixtures under rust-inoculated conditions. Similarly, varieties with a tendency to lodge benefited from being planted in mixtures because of support from more erect varieties.

As early as 1960, work done by the Institut National pour L'Etude Agronomique du Congo Belge (INEAC) demonstrated that a single variety, Caraotas, dominated the chosen mixture after four seasons of planting. Interestingly, researchers concluded that because of its undesirable seed type, its dissemination among farmers was not to be recommended; however, the experimental results were taken as evidence that mass dissemination of the best improved variety would eliminate the traditional mixtures (CIAT, 1986). Again, early conventional wisdom favored the pure "improved" line.

A major problem with all station research carried out on mixtures has been the exclusive use of synthetically created mixtures of station varieties. As a consequence, farmers' mixtures are often described as "land races," that is, a set of varieties that are more or less in a state of homeostasis. This assumption denies farmers' active management of their mixtures to optimize desired characteristics. A simple experiment—following actual farm mixtures for a few seasons without human intervention—could resolve this issue.

Table 2.1. Mixture preference

Prefer to sow mixture	Prefer to sow single variety	Mixture has better yield	Mixture more stable yield
96%	4%	61%	67%
N = 120			>100% multiple responses

Farmers' Management of Mixtures

Diagnostic Surveys

A diagnostic survey carried out by an anthropologist and a nutritionist among 120 farmers in Rwanda in 1984 demonstrated their overwhelming preference for planting mixtures. It also clearly demonstrated their understanding of the advantages of mixtures. The majority of farmers interviewed (96 percent) preferred planting mixtures because of higher stability of yield and higher overall yield (Table 2.1). The most common response to the question "Why do you grow mixtures?" was that some varieties would always produce something regardless of the seasonal conditions. Responsibility for the storage and management of seed turned out to be nearly 100 percent women's work (Table 2.2). The survey furthermore showed that farmers' mixture practices were more sophisticated than previously thought. Almost all farmers have different mixtures for different soil and intercropping conditions (Table 2.3). Typically, small-seeded mixtures were selected for less fertile soils, large-seeded mixtures for very fertile soil, and mixtures containing mostly upright-growth-habit plants for as-

Table 2.2. Bean production tasks by gender

	Prepare field	Sow	Choose seed	Weed	Harvest	Thresh	Seed storage
Always woman	82%	100%	100%	100%	88%	68%	100%
Sometimes woman	18%	—	—	—	12%	32%	—
Always man	32%	—	—	—	8%	36%	—
Sometimes man	64%	2%	—	10%	40%	50%	—
Never man	4%	98%	100%	90%	52%	14%	100%

Table 2.3. Number of different mixtures sown

	No. mixture				Percentage of farmers with a discrete mixture for each condition			
	1	2	3	4	Poor soil	Good soil	Banana association	Other
Percentage of farmers sowing mixtures	9%	37%	51%	3%	61%	65%	45%	14%
N = 120							>100% multiple responses	

sociating with bananas. Frequently, different varieties were selected depending on whether they were to be grown during the long or the short rains.

The reasons for this are straightforward: small-seeded varieties, although generally less liked, are known to produce under infertile conditions when large-seeded varieties no longer give an acceptable yield. Large-seeded varieties are preferred and are known to require a more fertile soil. Microclimatic conditions are more humid under bananas, and most of the beans produced during the heavy, long rains are grown under bananas. The use of upright well-aerated varieties reduces disease problems associated with humidity.

Of considerable concern to the project, since it was planned to initiate a series of on-farm varietal trials in collaboration with the national programs and development projects, was the question of whether new varieties should be tested pure or incorporated into farmers' mixtures. The survey showed that the majority of farmers (92 percent) experimented with new varieties, which they usually obtained from markets or from family and friends (Table 2.4). Farmers' own experimentation typically took the form of first trying the new variety pure on a small plot adjacent to the house. If it passed this test, the variety was then tried under various field conditions to determine when it produced an acceptable yield. Finally, sufficient seed was produced by again growing the variety pure and then incorporating it into the ap-

Table 2.4. Experimentation with new varieties

Never try	Try sometimes	Try often	Try separately	Try mixed	Save for seed if good	Multiply for seed if good
8%	52%	40%	78%	22%	96%	96%
N = 120						

Table 2.5. Follow-up of on-farm varietal adaptation trials by forty-five farmers after two to five seasons, 1986

Variety	Still grown	Seed given to other farmers	Sown mixed (M) or pure (P)	Cultivation conditions		
				Fertile soil	Infertile soil	Under bananas
Kiliumukwe	100%	51	P = 52% M = 48%	68%	4%	28%
Rubona 5	70%	24	P = 52% M = 48%	48%	17%	35%
Ikinimba	67%	24	P = 40% M = 60%	45%	45%	10%
Kirundo	65%	16	P = 34% M = 66%	72%	0	28%
A197	22%	0	—	—	—	—
Climbing mixture	27%	5	—	60%	0	40%

propriate mixture(s). Thus, a simple trial design that tests each variety pure corresponds with farmers' own practices. Follow-up studies of the trials were carried out to determine where in their cropping systems farmers commonly incorporated the varieties being tested.

Follow-up Surveys of On-Farm Trials

Follow-up surveys of the varietal trials were pretested in Rwanda in late 1985. In 1986, surveys were administered to forty-five farmers who had conducted on-farm trials two to five seasons previously. We wanted to know which of the tested varieties were still being grown, how and where they were grown, the extent to which farmers had diffused the acceptable varieties among their neighbors and kin, and whether they were still sowing the climbing beans they had received and, if so, how they had staked them and how they liked them. We also asked whether farmers tried new varieties pure or mixed when testing them for the first time and what their impressions of the trial itself were.

The results, summarized in Table 2.5, confirmed the high acceptability of the variety Kiliumukwe, still grown by 100 percent of the farmers. The real surprise was Ikinimba. In spite of its low overall evaluation, 57 percent of the farmers were still growing it because of its good performance on poor soils—a characteristic that had also been noted in the trial evaluations. The variety A197, which had performed quite respectably in the first trials, had the lowest level of retention at only 22 percent because of the yield instability over seasons resulting from its very high susceptibility to anthracnose.

Clear differences were found between varieties with regard to the extent to which they were being sown on rich soil, poor soil, or in association with bananas. Such data enabled the project to provide guidelines to projects and extension agents on where best to plant a certain variety.

The favored variety, Kiliumukwe, tended to be grown pure longer than the other varieties. This can best be understood as a process of seed multiplication. The preferred varieties are grown in pure stands longer in order to have enough seed to incorporate them at a higher proportion into the mixtures.

Differences in the spontaneous diffusion of varieties to other farmers were particularly striking. A total of 453 kg of Kiliumukwe had been given to fifty-one other farmers. The next best variety, Rubena 5, which was still being grown by 70 percent of the farmers, had approximately half this diffusion rate. It had been given to twenty-four others, and 270 kg of seed had gone out. Because of farmers' seed production and exchange practices, a relatively small amount of seed of a good variety will have a considerable impact if spread out thinly over the area where it is well adapted.

The majority of farmers were either content (25 percent) or very content (67 percent) with the trials. It was reconfirmed that farmers usually (85 percent) test new varieties pure when they are first experimenting with them. The validity of our approach in testing new varieties pure, rather than in mixtures, was thus corroborated.

Collection and Analysis of Mixtures. The diagnostic surveys did not prove very useful when it came to more detailed information about farmers' seed selection and mixtures management practices. For example, farmers frequently responded "no" when asked whether they changed the proportions of certain varieties between harvest and sowing. On the other hand, all farmers responded "yes" when asked whether they performed a "triage," or sorting, of their seed before sowing in order to eliminate "bad" seed. Direct observation of this sorting process showed that certain varieties seemed to be discriminated against, even if the seed did not appear to be blemished in any way. In order to obtain more detailed information on this seed-sorting process, farmers' mixtures were collected and analyzed from fifty farms across three agroecological zones (Crete, Plateau Central, and Mayaga) during 1986 and 1987.

At sowing time, farmers were asked to exchange a small amount of their seed for an equal amount of the variety Kiliumukwe (the highly preferred variety). Five kinds of mixtures were collected: unsorted beans directly from storage, seed sorted for sowing on very fertile soils, seed sorted for medium soils, seed sorted for infertile soils, and seeds rejected for sowing. The seed was collected directly from the field at the time of sowing. At this time the farmers were asked to classify the field according to its fertility. We then asked for samples of stored beans from which this seed had been derived and for the beans that had been rejected during the sort.

We compared the seed with the stored and rejected beans to measure what changes had taken place in the sorting process. Grain sizes were compared by passing the samples through a sieve, which allowed us to determine the percentage of large and small seeds in each sample. Similarly, the percentage of grains blemished with disease symptoms was calculated to determine the extent to which farmers cleaned up their seed prior to planting. The percentage of each of the main grain types in each sample was also calculated.

The results of this study provided more detailed information while confirming many of our earlier observations. For a sample of forty-two households, the mixtures varied between 6 and 29 varieties per mixture, with an average of 19.8 varieties. Ferguson and Sprecher (1987:11) reported an average of 12.9 varieties for Malawi. The difference may be due to the fact that in the Malawi study, varieties were differentiated by the farmers, whereas in our study this was done by researchers. Farmers do have a tendency to lump some similar varieties together under a common varietal name. Interestingly, station research carried out in Burundi (ISABU, 1983) showed that positive interactions among varieties in mixtures only occurred when the number of varieties in the mixtures was 6 or more. Farmers arrived at the same minimum number!

In spite of the large number of different varieties, three varieties (usually not the same three) accounted for 50–90 percent of each mixture. Seed from fertile fields was primarily large grained, while seed from infertile fields was mainly small grained.

The most useful comparison was among the rejects, the stored beans, and the actual seed. In all soil categories there was a considerably higher proportion of small grains among the rejects than among the chosen seed (Table 2.6). This clearly shows that farmers are compensating for the tendency of small-seeded varieties to take over the mixtures. For a given seed weight, there are many more grains for a small-

Table 2.6. Effects of farmer seed sorting on percentage of small- and large-grained beans

	Stored beans		Seed		Rejects	
	Small	Large	Small	Large	Small	Large
Fertile soil	40%	60%	30%	70%	77%	23%
Infertile soil	81.5%	18.5%	75.5%	24.5%	92%	8%

than for a large-seeded variety. This shows that researchers and extensionists need not be concerned about releasing varieties that are too competitive in the mixtures. The proportion of diseased seed was twice as high among the rejects as in the actual seed.

Still, there is considerable room for improvement of farmers' sorting practices, since 12.5 percent blemished grain was found in the seed stock. This last aggregate observation obscures the fact that there is considerable variability in the extent to which farmers clean up their seed during the sort. Some women do an excellent job, allowing virtually no blemished grains to pass through into their seed, whereas others do only a perfunctory sort. In this sphere, there is a great difference between knowledge and practice. Although almost all women know that a thorough sort is ideal, conflicting labor demands often mitigate against doing an ideal job.

We included seed with very slight disease blemishes in the category of "diseased seed." In practice, very few severely blemished grains are ever overlooked by the women and included in the seed. Fairhead (1990) found that one farmer criterion for deciding whether a blemish is acceptable is whether the coteledon is infected by the blemish.

Finally, we showed conclusively that farmers reduce the proportion of less desirable but higher-yielding small black grains while increasing the proportion of more desirable but lower-yielding yellow grain types. Of the rejects, 15 percent were small blacks, but only 2 percent were yellows. In the seed itself, 7 percent were small blacks and 14 percent were yellows. In this way, farmers keep the same relative proportions of these grain types over the seasons in spite of their yield differences.

Implications for Research and Technology Development

CIAT has consciously adopted a low-input approach which stresses natural genetic resistance over chemical solutions to many production constraints. This is especially so in the areas of disease and insect resistance and production under adverse soil and water conditions. This low-input approach based on genetic diversity is less destructive of the environment, socially more scale-neutral, and less expensive for farmers than chemical "packages" (Nickel, 1987).

The evidence presented above shows clearly that the use of bean cultivar mixtures provides a viable and inexpensive genetic solution to several of the main production constraints, particularly diseases, heterogeneous soils, and uncertain climatic conditions, facing most farmers in the Great Lakes region. It also shows that women, by preventing any single or a few varieties from taking over the mixtures, actively maintain a higher level of on-farm diversity than would be the case under natural selection.

Although the CIAT bean program has a clear interest in "broadening the genetic base" (CIAT, 1988), it may unintentionally have the opposite effect. Several times during the evaluation of on-farm variety trials it became clear that the technicians or extensionists responsible for the trials were extolling the virtues of planting the improved variety rather than the mixture. Our aim is to improve the composition of the mixtures by adding new varieties to them, not by replacing them. Research goals must be carefully explained to the people who provide the direct link with farmers so that messages will be passed on accurately.

Pressure on farmers to plant pure varieties often comes from the market. In Zaire, for example, whites and yellows command a considerably higher market price than complex mixtures. There are two ways of dealing with this problem. At national policy levels, state marketing and storage institutions can avoid the question of a multitiered price structure based on purity of grain type. In Rwanda, for example, the 1988 base price for beans fixed by the Office pour la Promotion, la Vente et l'Importation des Produits Agricoles (OPROVIA) is 35 francs per kilogram for all grain types, and pure lines are almost never seen in the markets. A positive national campaign stressing that "mixtures are beautiful" could be of great help in maintaining consumer acceptability and thus the market price of mixtures.

In some cases where price differentials may be difficult to tackle at a policy level, introducing more varieties of the desired color and diffusing them to farmers as mixtures can also increase stability while maintaining the higher rate of return for the farmer. In North Kivu, for example, the Programme Nationale des Legumineuses du Zaire, together with the CIAT regional program, is testing fifteen new white varieties. The aim is to introduce a mixture of the best of these varieties into the farmers' existing white mixture of usually only three varieties, thereby increasing its complexity up to six to ten varieties. Starting in September 1988, the same procedure was planned for the yellow-seeded group of varieties.

Since beans were introduced into Africa from Latin America, the diversity of African genotypes is considerably lower than that of Latin American genotypes.[3] One of the main things CIAT has to offer to African farmers and national programs is access to this source of diversity. Used properly, this can broaden the range of disease resistance and other desirable traits in the local mixtures, either through the successive introduction of new varieties into the mixtures or by improving existing varieties through backcrossing. Both approaches are currently being used.

Some acceptable local varieties already found in farmers' mixtures in some regions and currently being diffused throughout the region are being improved. For example, both the varieties Kiliumukwe in Rwanda and Kirundo in Burundi have very high farmer ratings; however, both are very susceptible to halo blight. A backcrossing program is currently under way to introduce halo blight resistance into these varieties while maintaining their other desirable characteristics, thereby improving the overall quality of the mixture.

A promising addition to this approach would be better use of the comprehensive bean collection survey carried out in Rwanda by Lamb et al. (1985). This survey identifies the varieties with the greatest geographical range and those planted by the greatest number of farmers—for example, the variety Carolina. Weaknesses in these varieties could be identified and corrected through a backcrossing program with both exogenous and local germ plasm. The improved local germ plasm could then be reintroduced as improved Carolina, for example.

A major project activity is the introduction of climbing beans into nonclimbing bean areas. Since climbing beans are much more productive on good soils than bush beans, this has long been identified as a promising avenue for the intensification of agriculture essential to offset population growth. The introduction of climbing beans into new

---

[3] Although there is clearly greater genetic diversity in the Latin American germ plasm, the African germ plasm's genetic potential should not be seen as simply a subset of the Latin American traits. Approximately one hundred years of natural and farmer selection, combined with a ± 2 percent outcrossing rate and natural mutations, have created varieties in Africa that have much to offer to the rest of the world. For example, to date, the best two nitrogen-fixing varieties identified by CIAT, Tostado and Carolina, both come from the Rwanda germ plasm collection.

areas presents a great opportunity for researchers to create synthetic mixtures that combine resistances to the major diseases and other complementary traits. Ferguson and Sprecher (1987:22) cited numerous sources[4] linking "the widespread introduction of new or improved cultivars with the loss of genetic diversity." There is a very real danger that this will happen with the climbing beans if they are introduced as single varieties.

Many researchers who have studied mixtures are now convinced of their great value in stabilizing and improving yields, especially under the low-input, suboptional conditions found in many developing countries. Still, a strong feeling remains that the practice of growing mixtures is ultimately doomed. I would contend that this is based on an often unconscious ethnocentric view that the future of agricultural development lies necessarily in emulating the practices in the more industrially developed nations. Ironically, in these countries there is now a move to replace single cereal varieties with mixtures or multi-lines because of their greater disease resistance.

A rapidly increasing sensitivity is developing to the fragility of the environment and the deleterious environmental and health effects of many agrochemicals. Along with intercropping and other cultural practices, genetic mixtures provide an important opportunity to achieve a more sustainable and environment-friendly form of agriculture in both developed and developing countries. Attaining this goal requires a reorientation of research objectives away from the breeders' holy grail of the supervariety and toward optimal forms of diversity achieved through combining farmers' knowledge about maintaining genetic diversity in the field with researchers' knowledge of genetic traits and improvements.

Bibliography

Abramovitz, Janet. *A Survey of U.S.-based Efforts to Research and Conserve Biological Diversity in Developing Countries.* New York: World Resources Institute, 1989.

Brown, W., T. Chang, M. Goodman, and Q. Jones, eds. *Conservation of Crop Germplasm: An International Perspective.* Madison, Wis.: Crop Science Society of America, CASSA Publication no. 8, 1983.

Chang, T. "Conservation of Rice Genetic Resources: Luxury or Necessity?" *Science,* 224:251–56, 1984.

[4] The cited sources are Brown et al., 1983; Chang, 1984; Frankel, 1970; Harlan, 1975; National Academy of Sciences, 1972; Wilkes, 1977; and Yeatman et al., 1984.

CIAT. *Bean Program Annual Reports, 1984–1988*. Cali, Colombia: CIAT.
——. *Bibliography of Bean Research in Africa*. Cali, Colombia: CIAT, 1987a.
Dessert, M. *Recherche sur haricot*. In *Rapport annuel*. Rubona, Rwanda: ISAR, 1986.
DeVos, P., K. Kabengele, and I. Nzimenya. "Legumineuses: Haricot *(Phaseolous vulgaris)*." In *Isabu rapport des activités de recherche*. Bujumbura, Burundi, 1983.
Edje, O. T., and M. W. Adams. "Stability of Bean Mixtures in Association with Maize." In *Bean Improvement Cooperative Annual Report*. Lilongwe, Malawi, 1985.
Fairhead, James. "Fields of Struggle: Towards a Social History of Farming Knowledge and Practice in a Bwisha Community, Kivu, Zaire." Ph.D. diss., University of London, School of Oriental and African Studies, 1990.
Ferguson, A., and S. Sprecher. *Women and Plant Genetic diversity: The Case of Beans in the Central Region of Malawi*. Bean Cowpea CRSP. East Lansing: Michigan State University Press, 1987.
Frankel, O. "Genetic Dangers of the Green Revolution." *World Agriculture*, 19:9–13, 1970.
Gahamanyi, Leopold. "Agricultural Research in Rwanda." *ISNAR Newsletter*. The Hague: ISNAR, 1985.
Harlan, J. "Our Vanishing Genetic Resources." *Science*, 188:618–21, 1975.
INEAC. "Haricots." In *Rapport annuel, 1960*. Rubona, Rwanda, 1960.
ISABU. *Rapport annuel*. Bujumbura, Burundi, 1983.
Ishabairu, T., and J. Teri. "The Effect of Bean Cultivar Mixtures on Disease Severity and Yield." *Phaseolus Beans Newsletter for Eastern Africa*, no. 2, 1984.
Jones, W., and R. Egli. "Farming Systems in Africa: The Great Lakes Highlands of Rwanda, Burundi and Zaire." Technical Paper no. 27. Washington, D.C.: World Bank, 1984.
Juma, Calestous. *Biological Diversity and Innovation: Conserving and Utilizing Genetic Resources in Kenya*. Nairobi: African Centre for Technology Studies, 1989.
Kloppenberg, J., and D. Kleinman. "The Plant Germplasm Controversy. Analyzing Empirically the Distribution of the World's Plant Genetic Resources." *Bioscience*, 37(3):190–98, 1987.
Lamb, E. M., et al. "A Survey of Bean Genotypes Grown in Rwanda." In *Bean Improvement Cooperative Annual Report 28*. Lilongwe, Malawi, 1985.
Lyimo, H. F., and J. M. Teri. "Effects of Bean Cultivar Mixtures on Disease Severity and Yield." In A. Minjas and M. Salema, eds., *Workshop on Bean Research in Tanzania*. Morogoro, Tanzania: Sokoine University Faculty of Agriculture, 1984.
Martin, G., and W. Adams. "Landraces of *Phaseolus vulgaris* F. in Northern Malawi. I: Regional Variation." *Economic Botany*, 41(2). New York: New York Botanical Garden, 1987.
McNeely, J., K. Miller, W. Reid, R. Mittermeier, and T. Werner. *Conserving the World's Biological Diversity*. World Bank: IUCN, WRI, CI, WWF-US, 1990.

McPherson, Malcolm. "Critical Assessment of the Value of and Concern for the Maintenance of Biological Diversity." Development Discussion Paper 212, Harvard Institute for International Development, 1985.

National Academy of Sciences. *Genetic Vulnerability of Major Crops.* Washington, D.C.: National Academy of Sciences Press, 1972.

Neeley, W., W. McProud, and J. Yohe. "Diversity by Breeding: Genetic Variability on the Farmer's Field." Paper presented at the Agricultural Development Conference, Risk and Uncertainty in Agricultural Development. CIMMYT, Mexico, 1976.

Nickel, John L. *Low-Input, Environmentally Sensitive Technologies for Agriculture.* Cali, Colombia: CIAT, 1987.

Sirven, P., J. F. Gotanegre, and C. Prioul. *Géographie du Rwanda.* Brussels: De Boeck, 1974.

Trutmann, P., and P. Mukeshi. "Effects of Bean Mixtures on the Spread and Severity of Angular Leaf Spot." In *CIAT Bean Program Annual Report.* Cali, Columbia: CIAT, 1988.

Wilkes, G. "The World's Crop Plant Germplasm—An Endangered Resource." *Bulletin of the Atomic Sciences,* 33:8–16, 1977.

Yeatman, C., D. Kafton, and G. Wilkes. "Plant Genetic Resources. A Conservation Imperative." Washington, D.C.: American Association for the Advancement of Science Symposium 87, 1984.

3

"The Friendly Potato":
Farmer Selection of Potato Varieties
for Multiple Uses

Gordon Prain, Fulgencio Uribe,
and Urs Scheidegger

Introduction

Why is it that rural people's knowledge—such an enormous and rich resource in poor countries—has been largely ignored by those countries' governments? Ignorance of farmers' own investigations into the local agricultural system can increase official research costs, delay the introduction of innovations, and in some cases lead to inappropriate development interventions. One of a number of efforts to incorporate farmers' experience and knowledge into the research process[1] was undertaken at the International Potato Center (CIP) in the late 1970s and early 1980s, resulting in the formulation of the "farmer-back-to-farmer" model of development research. The farmer works alongside the researcher in identifying the problem and working out possible solutions, and in the subsequent technology testing and adaptation. If the technology is neither adopted nor adapted, both farmer and scientist go back to the drawing board (Rhoades and Booth, 1982).

One of the great advantages of this model, paradoxically, is that it

The authors benefited from comments on an earlier draft of this paper by Charles Crissman, Douglas Horton, Joyce Moock, and Robert Rhoades.

[1] See the recent review of farmer participation in agricultural research in Farrington and Martin, 1987.

allows the research focus to go beyond agricultural production in a way that farming systems research has not always managed to do. The original CIP research that led to the farmer-back-to-farmer model involved seed potato storage. More recent research has involved the study of peasant seed systems, which include household seed management, production practices, and a whole set of exchange relations linking households and markets. The model builds on the fact that most rural people who farm also store and eat the food they produce and use by-products as feed; many also process it and exchange it in a market.

As a consequence of this comprehensive involvement in the food system, farmers are assiduous if opportunistic plant breeders; they evaluate food plants as crops, as sources of family nutrition, and as commodities. The existence of several hundred native potato varieties with a wide range of characteristics in the Peruvian Andes is due to the careful selection and subsequent evaluation by farmers of naturally occurring crosses.

Official germ plasm improvement programs have often lacked this broad farmer perspective, concentrating instead on narrower production considerations, especially improved yield. Many of the new rice varieties, for example, were selected for high yields, but stalk length, a crucial source of forage during the dry season, is severely reduced. Selections are also frequently made on research stations with near-optimum conditions—very different from the circumstances of most small farmers. Where breeders are familiar with peasant farming, their secondary selection criteria may coincide with the farmers', but would it not be better to incorporate farmer criteria into the breeding process itself?

To understand the alternative perspective that farmers can give to breeding programs, we introduce the concept of "the friendly potato." This is a rough translation of the Spanish *la papa simpática*. The term *simpático* is usually applied to a person and describes someone who is obliging, understanding, adaptable, one of the gang—in general, someone you like to have around. When applied to a potato, it seems to capture very well what farmers look for in a new variety— it adapts easily to the farmers' food system. The term was coined by one of the farmers who participated in a series of variety trials organized by the Peruvian National Potato Seed Program during the 1987–88 season. This chapter presents some of the results of those trials and in so doing attempts to spell out in some detail what it means for a potato variety to be *simpática*.

Background and Methods

The Peruvian Andes are the center of genetic diversity of the potato (Hawkes, 1978). The potato accounts for over 60 percent of the economic value of the fifteen major highland crops and is the most important ingredient in the highland diet (Fano, 1986). Potatoes are cultivated in diverse mountain ecologies. They are mainly grown on the steep sides of intermontane valleys and in depressions and ravines of the *puna*—high-altitude lands that separate these valleys from the desert coast to the west and the lowland jungle to the east. They are also grown in many valleys that are directly open to the warm, humid air currents of the jungle. The distinct physical environments in which potatoes are produced cause different problems and needs. With such vertical and horizontal variation, the strategy of Andean farmers has always been to cultivate in many different ecological zones, using many different genotypes, so that it is common for households to have a "farm" of fifteen or twenty small parcels of land located at different altitudes and in different microenvironments (Murra, 1986). The extent to which such diversification is possible also strongly influences the economic roles potatoes play for different households, and both ecological and economic considerations influence the varieties farmers select (Prain and Uribe, 1986).

Systematic breeding work in the Peruvian highlands began in 1947 under the direction of Carlos Ochoa (Ochoa, 1961). The dominant breeding criteria then were "yield, rusticity, resistance to late blight, frost and resistance to potato wart *(Synchytrium endobioticum)*" (ibid.:2). Selections were made under a range of conditions and at altitudes between 3300 and 3900 m above sea level. Two varieties were released during the 1950s and widely adopted by farmers (ibid.:11).

More recent breeding work in the central highlands has been carried out under narrower environmental conditions, mainly in the experiment station in the flatter, more fertile floor of the Mantaro Valley. These conditions are similar to those enjoyed by a small minority of large commercial producers who have been the main beneficiaries of the varieties released (Chambers and Ghildyal, 1985; Franco et al., 1979; Prain and Scheidegger, 1988:189).

Peru has also initiated several programs to produce high-quality seed of the new varieties, most recently in the 1960s (Ezeta and Scheidegger, 1985; Ministerio de Agricultura, n.d.) These programs were not generally successful, mainly because the "improved" seed was not very good and the farmers' seed was not as bad as had been thought (Horton, 1984:8). A new seed program was started in 1983 using modern techniques of heat therapy and rapid multiplication (Bryan et al., 1981) to produce virus-free seed for distribution to farmers (Ezeta and Scheidegger, 1985).

The program has so far mainly involved working with modern and native varieties already known and used in particular regions. But there is also a need for the seed program to link up with the breeding program to facilitate the introduction of new genetic material that more successfully meets particular needs. The farmer expertise already identified in surveys and in other trials (Scheidegger et al., 1989) could be tapped in order to identify appropriate germ plasm.

Five locations in the Mantaro, Cunas, Yanamarca valleys were selected, with altitudes ranging between 3550 and 4000 m above sea level. There was no attempt to cover the whole range of ecological and economic diversity. We did not include, for example, the large commercial producers who control 10–15 percent of the area planted to potatoes, since the program was mainly concerned with addressing the needs of small producers. Apart from the restriction to peasant farmers, the small amount of planting material available limited the number of locations that could be chosen and the number of farmers who could participate. The same set of clones was planted in each location. It was possible to plant only an incomplete set of clones in the highland experimental station located in the Mantaro Valley bottom (3250 m above sea level) for comparison because of the shortage of planting material. To increase farmer input and stimulate the swapping of opinions and experiences, we divided the clones into two groups (Table 3.1) given to two farmers in each location. Cooperators were selected from among the large number of farmers with whom good working relationships had been established in the previous work of the program on the basis of their particular interest in potato varieties. In one case, a small neighborhood group within a village offered to carry out the selection work jointly. This particular site is a high-altitude zone that suffers severe frosts, and the crop is grown mainly for subsistence. Under such circumstances, individual participation in experiments is difficult and the group involvement offered a wide impact with little individual risk.

Most of the material (Table 3.1) came from two CIP breeding programs: the nematode *(Globadera pallida)* program and the frost and late blight *(Phytophthora infestans)* program; some further clones were

Table 3.1. List of clones used in farmer selection trials, 1987–88

CIP code	INIPA code	Trial name
	Group 1	
G3		Tumbes
HFF20.2		Chota
UFF12.2		Huaral
		Chejche (Bolivian native variety)
		Caprio (Colombian variety)
	S-24-73	Chuco
	S-229-72	Yauri
		Yungay (control)
	Group 2	
683246.12		Huaraz
280179.7		Tacna
375597.15		Huanta
HFF4.2		Huacho
UFF4.2		Yauyos
P3		Perricholi (Peruvian variety, just released)
	PI-15-19	Chavin
	PI-29-19	Aija
	PI-2-8	Mayocc
		Yungay (control)

provided by the Peruvian National Potato-Breeding Program. A Colombian variety and a Bolivian native variety were also included. A name was given to each clone to facilitate discussions about the different clones and their characteristics. A successful modern variety, Yungay, which is widely distributed in the central highlands due to its "rustic" characteristics, was included in both sets of clones as a control. It was assumed the clones would need to show some notable additional advantages over Yungay to be of interest to farmers.

Farmers were responsible for selecting fields and designing the trials. In most cases they chose to plant all fifty tubers of each clone in one row. Foliage development at these altitudes is limited and soil cover normally does not surpass 70 percent; thus little intervariety competition is to be expected. Farmers used their own seed as a border around the trial and to fill up the field. In some cases the plot was too narrow and each clone was planted in two adjoining rows. No "spatial" replications were laid out, but in farmer experimentation, "temporal" replication is considered a more relevant way of dealing with variation (see p. 67).

The team was present at all plantings and afterward visited each of the five sites every two to three weeks to discuss the performance of clones during the growing cycle. Each farmer was given a notebook to record details of the trials and observations about characteristics. In addition, we also recorded farmers' comments. We were present at each harvest to weigh output and record the family's observations. In most cases we were able to arrange for the two farmers with neighboring trials to exchange visits to look at the different group of clones, stimulate further comments, and arrange exchange of clones after the harvest.

Three "group evaluation sessions" were organized for all farmer cooperators (with two representatives from the neighborhood group) to encourage joint discussions, exchange of opinions, and expression of priorities. The first session, in Quicha Chico, evaluated the two trials during the growing season; the second evaluated the two harvests in Marcavalle. The third session was held in the experimental station where the program office is located after all the harvests were finished. In this last session, all the yield results were displayed and our own summary of farmers' comments made during the session was available as a memory aid.

Results

Yield Evaluation

There is a clear "yield gap" between the optimum conditions of the station and the real conditions of farmers' fields (Figure 3.1). More interesting, however, are the genotype-environment interactions. The native variety Chejche and the clone UFF4.1 might have been weeded out on the basis of poor station yields, but they performed reasonably well for farmers. Conversely, very good station yielders like 375597.15 and P3 did not perform spectacularly for farmers. The clone 280179.7 was one of the better yielders on-station but was the second poorest performer in the farmer trials. It was firmly rejected by two of the four farmers who planted it because of its small, deformed, "characterless" tubers. These results suggest that genotypes selected for yield under optimum conditions may not be the ones farmers would choose.

It is also of interest to look at yield stability across the different trial sites, since this is an important characteristic in an environment as diverse as the Andes. "Yield stability analysis" compares the yield of each clone across different sites with the "environmental index" of each site. The variety Yungay, which is widely distributed in the cen-

Figure 3.1. Average farmer yields and experiment station yields

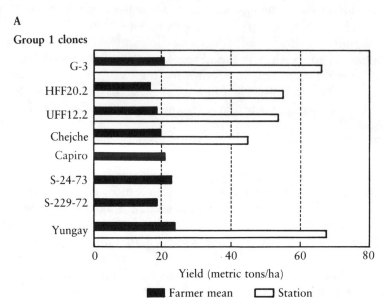

A
Group 1 clones

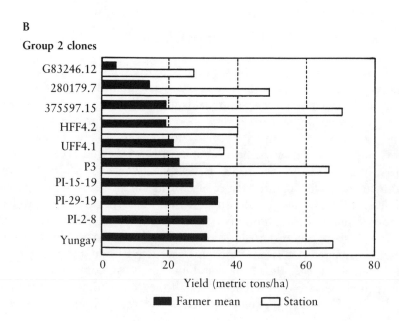

B
Group 2 clones

tral highlands, was taken as the environmental index against which to plot the yields of the other genotypes (Figure 3.2). The shallower the slope, the more stable the clone. Clones below the diagonal yielded less than Yungay. P3 is a highly unstable yielder on-farm compared with UFF4.1 (Figure 3.2D), which had a similar average yield (Figure 3.1A). The two genotypes' results on-station would have favored the selection of P3.

The native variety Chejche is a stable if modest yielder (Figure 3.2B) and was popular with farmers. G-3 has similar stability to Yungay and yields less, but it was judged better than Yungay by most farmers.

While these yield results show how necessary on-farm selection trials are, it is also clear that there is a need to look beyond yield and yield stability data at other factors that farmers consider important.

Food System Evaluation

Farmers spontaneously evaluated varieties using thirty-nine criteria, which fall into seven distinct categories (Table 3.2).

Physiological and agronomic: Observations were collected relating to foliage volume, stem number, and stem thickness, which were often discussed generally as "good development" or "little development" of the stand. A variety with a little stand was a cause for some concern, and in one case was fed some extra urea at hilling-up time. But all farmers agreed that what counts is what is below the ground. All had experienced excellent-looking stands that did not produce tubers. One of the most popular varieties at the end of the trial, G-3, had one of the smallest, weakest stands in all experiments. As one farmer said when she harvested a large number of good-sized tubers of this variety, "the worst is now the best."

Earliness, which is evaluated in terms of velocity of emergence and velocity of foliage development as well as maturation time, is an important characteristic for farmers, especially for its contribution to household nutrition. Early varieties are desirable to fill seasonal scarcities in food availability. In most of the highlands, rain falls from about October to March or April, and since the vast majority of peasant farmers have no irrigation, this is the growing season, with main harvesting taking place in May or June. Severe frosts between May and late September also make it risky to attempt earlier plantings in the higher areas. With these climatic constraints, farmers try to grow varieties with different vegetative periods. Early varieties relieve the boredom of a constant diet of wrinkled potatoes, which is all that is

Figure 3.2. Yield stability of clones compared with Yungay

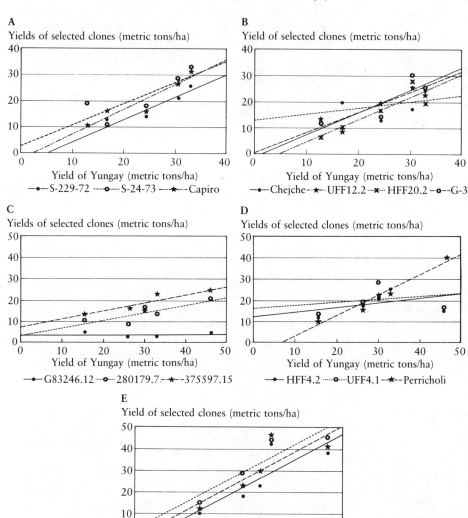

available of the previous season's harvest from January onward. Farmers agreed that it is acceptable to produce less if it is earlier. That was why the national program clone PI-15-19, though not the highest yielder, was of special interest. Earliness also has important economic implications if the household's need for new food has already been met. These varieties can catch the higher off-season prices in the markets and bring needed cash into the household.

Other agronomic evaluations with economic implications concerned uniformity of emergence and stolon length. Uniform emergence of plants contributes to easier, more efficient cultural practices. Stolon length affects the efficiency of the harvest. If the tubers are "laid like a hen's eggs," close to the plant, harvesting will be quicker and less costly. Labor is released for other tasks—very important at a time of constant labor shortage—and more of the production gets into the sack.

Performance in environment: One physiological characteristic with environmental implications is leaf size. Several trials suffered hailstorms, and Capiro and UFF 12.2 were both identified as surviving the hail better because of smaller leaves. Farmers were also interested in frost resistance and recuperation from its effects. However, there was little serious frost during the 1987–88 season, and a further season was considered important to evaluate this criterion.

Pests and diseases: Pest damage tended to be observed mainly during the harvest rather than preharvest, with special attention given to the Andean weevil. However, one farmer noted in the preharvest period that PI-2-8 appeared to resist a serious attack of *Epitrix* better than the other varieties.

Yield and size distribution: Although harvest yield was sometimes assessed in general terms as "very good," "reasonable," and "little," evaluations were usually more specific, relating to what farmers call the "quality" (calidad) of the yield, which means the percentage of first-size, better-priced tubers produced. Since most farmers grow modern varieties primarily as a cash crop (Table 3.3), "quality" in this special sense assumed major importance in these trials.[2] For these

[2] Farmer-experimenters generally manage two or three different classes of potato varieties. "Modern" refers to the varieties developed as part of the national breeding effort, "native commercial" to indigenous varieties that are individually selected, often for their commercial potential, and have names known over a wide area. Their main culinary and commercial characteristic is their floury texture, a characteristic that commands a higher price. "Gift varieties" *(papas regalos)* comprise a large range of indigenous cultigens that are grown in mixed plantings and are almost exclusively for subsistence. They are an im-

Table 3.2. Criteria mentioned by farmers participating in the evaluation of clones and new varieties of potatoes

Criteria	Desired level	Implication/explanation	Importance subs/mark[a]
Physiological-agronomic			
Velocity of emergence	High	Indication of earliness	1/1
Uniformity of emergence	High	Facilitates timing of agricultural practices	1/1
Velocity of foliage development	High	Indication of earliness	1/1
Maximum foliage volume	High but not abundant	Better yield possibility and good harvest index	2/2
Stem number	Medium	Similarity to native varieties	2/2
Stem thickness	High	Less lodging	1/1
Leaf size	Small	Better hail resistance	2/2
Stolon length	Short	Easier to harvest	2/2
Earliness	Options	Early varieties for first food and early season prices, late varieties for high yields in main harvest	2/2
Performance in environment			
Resistance to frost	High	Better yield	2/2
Recuperation from frost	Rapid	Better yield	2/2
Resistance to hail	High	Better yield	2/2
Recuperation from hail	High	Better yield	2/2
Pests and diseases			
Resistance to Andean weevil	High	Andean weevil is the most important pest, but no good resistance was present in the material	3/3
Resistance to *Epitrix sp.*	High	Lower spraying costs	1/1
Resistance to *Phoma andina*[b]	High	Lower spraying costs	1/1
Number of rotten tubers	Low	Lower spraying costs	2/2
Incidence of powdery scab	Low	Better appearance for marketing	1/2
Yield and size distribution			
Yield[c]	High		2/2
Percentage of first-size tubers (*calidad*)	High	Better price, easy to harvest	2/3
Total numbers of tubers	Medium	Enough seed for multiplication but also first size	2/2
Percentage of small tubers	Medium	Shorter harvesting time	1/1
Tuber characteristics[d]			
Skin color (primary/secondary)	Strong colored	Easier marketing, better price	1/3
Skin anomalies	Absent	Easier marketing	1/2
Tuber shape	Options	Either native-shaped or modern-looking with shallow eyes	1/2
Percentage of deformed tubers (cracks, secondary growth)	Low	Easier marketing	2/3
Uniformity in tuber shape	High	Varietal recognition	2/2
Fresh color	Yellow	Indication of culinary quality	2/2
"Wateriness" (ease of squeezing out juice)	Low	Indication of culinary quality	2/2

Criteria	Desired level	Implication/explanation	Importance subs/mark[a]
Cooking and eating quality			
Cooking time	Short	Less fuel consumption	2/1
Skin splitting at boiling[e]	Options	Indication of culinary quality	1/1
Ease of peeling (raw/boiled)	Easy	Soups/village processing	2/2
Texture	Flowery or sticky	Both for boiled potatoes	3/1
Taste	Not quantifiable	Must have distinctive taste	3/1
Storage behavior[f]			
Dormancy period and sprout growth	Long, slow	Maintain quantity of seed and consumer stock	2/2
Water loss	Low	Maintain weight for marketing	1/3
Tuber rotting	Low	More produce	2/2
Appearance after storage	Not wrinkled	For marketing	1/3

[a] Importance given to this criterion by farmers with respect to varieties destined for subsistence ("subs") and for marketing ("mark"); 1 = low, 3 = very high importance.

[b] Farmers refer to late blight *(Phytophthora infestans)* and phoma leaf spot *(Phoma andina)* with the same term, *rancha*. Since in the agroecological level referred to here late blight symptoms in the foliage are not easily seen, and since farmers pointed out phoma leaf spot symptoms, *rancha* is translated as *"Phoma andina"* in the case of foliage diseases. On the other hand, when *rancha* refers to rotten tubers it is translated as *"Phytophthora infestans,"* since *Phoma andina* does not affect tubers.

[c] Yield, deformities, and other tuber characteristics are evaluated by farmers in relation to soil fertility and depth, soil color (content of organic matter), quality of land preparation, previous crop, and altitude. While farmers prefer varieties that perform well in all soils and at different altitudes, they are willing to accept varieties with narrower agroecological adaptation.

[d] The whole set of tuber characteristics is synthesized by farmers in the term *friendliness (papa simpática).*

[e] Farmers consider varieties that split after normal boiling as of better culinary quality. However, as farmers usually boil a mixture of varieties for their own consumption, a certain degree of boiling consistency should be maintained, and therefore varieties that split after a very short cooking time are not acceptable.

[f] Only random data on storability could be obtained because of the worsening security situation. We include here information from other work in the same areas.

varieties, in both highland and Lima markets, there is at least a 60 percent price difference between second- and first-size potatoes. When supplies are abundant, this can rise to over 100 percent.

Several clones had unspectacular quality (Table 3.4). They received few comments, often contradictory, indicating unstable performance across different environments. Others, such as G83246.12, were poor across all of the trials, which together with their poor yield suggests

---

portant element in exchanges between farming households and have individual names usually known only within a local area.

Table 3.3. Importance of different potato varieties among farmer-experimenters

Location	Farmer	Modern[a]	Native commercial[b]	Gift[c]
			Type of varieties	
Marcavalle	1	S	V	V
	2	U	V	V
Chicche–La Libertad	1	V	V	V
	2	V	V	S
Cunas	1	S	V	U
	2	V	S	U
Quicha Chico	1	V	V	S
	2	V	V	S
Acolla	1	V	S	U
	2	S	V	U

Note: U = unimportant; S = somewhat important; V = very important.
[a] Sales, food (soups), chuño.
[b] Sales, food (boiled), chuño.
[c] Food (boiled), chuño.

Table 3.4. Frequencies of "quality"[a] ratings for clones in selection trials

Clone	Good	Poor	Total
		Rating	
Group 1			
G-3	3	0	3
HFF20.2	1	0	1
UFF12.2	2	1	3
Chejche	1	1	2
Capiro	0	3	3
S-24-73	2	0	2
S-229-72	0	1	1
Yungay	2	1	3
Group 2			
G83246.12	0	4	4
280119.7	2	2	4
375597.15	2	1	3
HFF4.2	1	1	2
UFF4.1	2	0	2
Perricholi (P3)	3	0	3
PI-15-19	0	2	2
PI-2-8	1	0	1
Yungay	3	1	4

[a] *Quality* is a translation of the Spanish term *calidad* used by farmers to refer to the percentage of first-size tubers.

that they have no future in farmers' fields. Capiro came out very poor in quality but yielded reasonably well, producing a large number of small tubers. It had, however, some additional characteristics that varieties with poor quality need to show to be cultivated. Apart from its hail resistance, earlier farmers also commented on its floury taste.

Perricholi showed the opposite tendency. It had a generally good quality rating, but an equal number of farmers noted that there were very few tubers. This causes problems when the farmer comes to select seed for the following year's planting. Varieties must produce an adequate quantity of small tubers for seed, since self-sufficiency in seed is the basis for Andean potato farming.

The farmers were particularly happy with the clone G-3, which produced good-sized tubers sufficient in number to allow the maintenance of the seed stock.

Tuber characteristics: Apart from the size and number of tubers, farmers also commented on skin and flesh coloring and form. Color is another feature that affects marketability, since colored potatoes usually fetch a higher price than white ones. This was an additional attraction of G-3, whose bright red color was variously described as "pretty," "lovely," and, more circumspectly, "interesting." Its elongated shape also provoked comment. One farmer described it as "like a bull's horn," a characteristic typical of native varieties. Even the colored skin of Capiro to some extent compensated for its lack of quality according to some farmers. The two other colored clones— G83246.12 and G280119.7—had too many other disadvantages to be of interest: the former for its miserable yields and the latter because of the large number of deformed tubers and the lack of a typical shape.

Farmers were concerned about tuber deformities and anomalies in skin texture because of their effect on marketing. More generally, they considered it important both for the market and for subsistence that a clone have a uniform shape so as to permit its clear identification as a variety.

Cooking and eating quality: The information on kitchen performance was collected during the weeks after harvest, after each household had had a chance to try out all the clones they had harvested. Cooking time of particular food crops and varieties is of importance in these high-altitude regions where wood is scarce and kerosene is expensive and difficult to obtain. The clones PI-2-8 and G-3 and the control variety Yungay were said to be difficult to cook.

Farmers had many comments on taste and texture of different varieties. Taste was evaluated in general terms as "pleasant" or "unpleasant," whereas texture was more specific, describing the "flouriness," "stickiness," or "wateriness" of particular varieties. PI-15-19 got the highest negative score as unpleasant and watery, and the native variety Chejche received the highest positive score. S-24-73, 375597.15, and Capiro were also appreciated by several farmers.

Storage behavior: No systematic data were collected about storage behavior of the different clones because of the worsening security situation. Table 3.2 includes information obtained during previous work in the same areas.

Conclusions

The first conclusion from the results of these trials is that the yield of a variety under optimum conditions is a very poor indicator of its likely adaptability or acceptability. Breeders are generally aware of this problem and attempt to test their clones under a range of environments. But we have also seen that while the adaptability of a clone can be better evaluated through testing under variable *farm* conditions, acceptability depends on a wide range of *farmer* evaluations. However, peasant farmers are not looking for the ideal variety. The evidence suggests that farmers seek to manage an ideal *range* of varieties which answers their food system needs (Prain and Scheidegger 1988:190). Individual varieties are selected in terms of their fit with diverse ecological conditions and diverse uses. They are often allocated specific ecological niches where positive characteristics flourish and negative aspects least express themselves. They are evaluated in terms of two main uses: as a cash crop and as a household food. Although some varieties satisfy one requirement more than the other, there is considerable overlap; some preoccupation with the taste of varieties mainly destined for the market suggests that all varieties may be required to feed the family in times of hardship. A "friendly potato" is a variety that satisfies one or other requirement while maintaining this flexibility.

Two friendly potato candidates emerged from the postharvest group evaluation held in the experimental station. Clone G-3 mainly satisfies market requirements. It is late emerging and has a weak stand, but tubers are formed close to the plant, facilitating the harvest; it performs well in the area; hailstorms do not seriously affect it; it produces a reasonable number of large tubers with enough smaller ones for seed; and it has an attractive color and shape for the market. It is better for use in soups than for eating boiled. Clone S-24-73 mainly satisfies household needs. It has rapid emergence and matures quickly; it has a strong stand with many stems; it can stand up to light frost but may be a little susceptible to hail; it produces plenty of good-sized tubers; it has good-shaped, white-skinned tubers that "look like a native variety"; and it cooks quickly and tastes good.

Even though these clones were appreciated by the farmers, they were not ready to endorse their release as varieties. We discovered that farmers have a long-term view of variety selection which is based on a detailed knowledge of ecological and climatic variation. One farmer commented on the very light frosts that season and the need to subject the varieties to a stiffer climatic test. All farmers wanted to see how the varieties performed in different soils and rainfall patterns. We realized that although no farmer will ever suggest laying out an experiment with four replications side by side, farmer replications do occur—over time and in different seasons. A further year of experiments will continue with minimum program input.

The complexity of both the local ecology and the household economy is amply reflected in the thirty-nine criteria that farmers consider in their evaluation of varieties. Breeders can never hope to satisfy this diversity with one variety. Not only are the numbers daunting, but in some cases the actual desired level of a particular criterion is variable. In the case of earliness, for example, there are two desired options: short vegetative period to overcome seasonal food scarcity and to take advantage of higher market prices, and a longer vegetative period for obtaining maximum "quality" and yield in the main harvest. What breeders can do is recognize the need for a range of varieties that satisfy different segments of the farmers' food system.

Beyond the first major selections of the breeding pipeline, farmers are best qualified to evaluate how far the new material is "friendly" to the different segments of their system. By involving farmers as partners, breeding programs have the chance to be more cost-effective in the development and release of varieties well adapted to local and national food systems.

Bibliography

Bryan, J., M. Jackson, and N. Melendez. *Rapid Multiplication Techniques for Potatoes.* Lima, Peru: International Potato Center, 1981.

Chambers, R., and B. P. Ghildyal. "Agricultural Research for Resource-Poor Farmers: The Farmer-First-and-Last Model." *Agricultural Administration,* 20(1):1–30, 1985.

Ezeta, F., and U. C. Scheidegger. "Basic Seed: A New Production and Distribution Program for Peru." *CIP Circular,* 13(2), 1985.

Fano, Hugo. *Los cultivos transitorios en el Perú: Análisis de su estructura y tendencias de 1964 a 1979.* Lima: CIP, 1986.

Farrington, John, and Adrienne Martin. "Farmer Participatory Research: A Review of Concepts and Practices." Agricultural Administration (Research and Extension) Network Discussion Paper 19. London: Overseas Development Institute, 1987.

Franco, E., D. Horton, and F. Tardieu. *Producción de la papa en el Valle del Mantaro, Perú. Resultados de la Encuesta Agro-Económica de Visita Unica.* Departamento de Ciencias Sociales. Documento de Trabajo 1979-1. Lima: CIP, 1979; reprint 1981.

Hawkes, J. G. *History of the Potato.* In P. M. Harris, ed., *The Potato Crop.* London: Chapman and Hall, 1978.

Horton, D. *Social Scientists in Agricultural Research. Lessons from the Mantaro Valley Project, Peru.* Ottawa: IDRC, 1984.

Ministerio de Agricultura (Peru). *Proyecto de incremento de la producción de papa 1967–1970.* Lima: CIPA, n.d.

Murra, John. "The Economic Organisation of the Inca." *Research in Economic Anthropology Supplement,* ser. no. 1, 1986.

Ochoa, Carlos. "Selección de hibridos de papa." *Revista Agronomía,* 28(34):1–16, 139–42, 1961.

Prain, G., and U. Scheidegger. "User-Friendly Seed Programs." In *The Social Sciences at CIP.* Report of the Third Social Science Planning Conference held at CIP, 1987. Lima: International Potato Center, 1988.

Prain, G., and F. Uribe. *El conocimiento campesino en la cosecha, selección y classificación de papas.* Minka 20. Huancayo, Peru, 1986.

Rhoades, R., and R. Booth. "Farmer-Back-to-Farmer: A Model for Generating Acceptable Agricultural Technology." *Agricultural Administration,* 11:127–37, 1982.

Scheidegger, U.C., G. Prain, F. Ezeta, and C. Vittorelli. "Linking Formal R&D to Indigenous Systems: A User-oriented Potato Seed Programme for Peru." Agricultural Administration (Research and Extension) Network Paper 10. London: Overseas Development Institute 1989.

4

Farmer Knowledge and Sustainability in Rice-Farming Systems: Blending Science and Indigenous Innovation

Sam Fujisaka

Introduction

The Green Revolution in Asia—where 90 percent of the world's rice is grown and eaten—resulted from the introduction and widespread adoption of semidwarf, high-yielding varieties of rice in the late 1960s and 1970s. For these areas, the post–Green Revolution period has been characterized by a slowing of the rate of growth of farm yields, high input-use intensity, and an apparent decreasing efficiency of inputs. In the post–Green Revolution era more attention has been given to nonirrigated rice environments, which account for about 50 percent of total rice area and 25 percent of production. Nonirrigated areas are more adverse environments than irrigated ones. Drought, flooding, and adverse soils contribute to low and unstable yields in about two-thirds of the rainfed lowland rice area, while infertile, usually acid, soils, weeds, drought, and soil erosion have led to low, unstable, and, in some areas, declining yields in upland rice.

Overall, the slowing yield growth rates and decreasing input efficiency in irrigated rice, and low and declining yields in nonirrigated rice, have meant that the sustainability of rice production systems has been called into question. At the same time, resource-poor farmers and landless people—who now consume more than half of all rice

Table 4.1. Some farmer technologies and places of origin

Farmer technologies[a]	Some places of origin
Rotary push weeder, variations	Japan, Indonesia, Sri Lanka
Five-tined furrow opener, weeder	Philippines, India, Myanmar
Intercropping, multiple cropping	China, Japan, Indonesia, Sri Lanka, Philippines
Azolla	China, India, Vietnam
Green manures	China, Korea, Japan, Philippines, India, Bangladesh
Fertilizer balls	China, Japan
Soil dressing (upland soil in paddy)	Madagascar
Dry-seeded rice	Bangladesh, Indonesia, Philippines, India, Myanmar
Beusani	Nepal, India, Bangladesh
Dapog seedbeds	Philippines
Rice–upland crop *sorjhan*	Indonesia
Ratooning for fodder	Madagascar

Source: Partially adapted from Goodell, n.d.
[a] Not included are terracing, bunding, puddling, drainage, or flooding for pest control, syncronous planting, and other practices that are widespread and traditional.

produced—may number more than 1 billion shortly after the turn of the century.

Post–Green Revolution agricultural science is realizing the need to incorporate the previously underutilized resources of farmer experimentation (Box, 1987; Rhoades, 1987), farmer participation in formal research (Ashby, 1987; Biggs, 1988; Farrington and Martin, 1987) and technology development (Matlon et al., 1984; Sagar and Farrington, 1988), and indigenous technical knowledge (Brokensha et al., 1980; Richards, 1985; Thurston, 1990; Warren and Cashman, 1988) in the development of technologies that enhance sustainability (Table 4.1). "Farmer-back-to-farmer" (Rhoades and Booth, 1982), "farmer first" (Chambers et al., 1989), and other terms have been coined to describe these new farmer-scientist research partnerships.

Green Revolution rice research—as conducted by the International Rice Research Institute (IRRI)—has been popularly associated with irrigated environments, reduced genetic variability, and chemical input use, rather than with the farmer-back-to-farmer approach. Even formal Green Revolution research, however, built upon farmers' practices and knowledge. Because rice scientists did not openly acknowledge such farmer inputs, the approach remained implicit and unsystematized. Current research at IRRI in nonirrigated areas and in the irrigated areas facing declining sustainability is attempting to recognize, systematize, and improve the process of incorporating farmers' knowledge into research. This chapter presents cases that illustrate how farmer science and formal science can be complementary in the development of more sustainable rice systems.

Green Revolution Farmers and Formal Science

Agricultural engineers have long worked to improve traditional implements. The now widespread push-type hand weeders originated from implements such as the nail weeder, Indian rotary peg weeder, and Indonesian and Japanese single-row rotary weeders (King, 1911; RNAM, 1983). Animal-drawn interrow cultivators, treadle threshers, animal-driven pumps, and other "IRRI technologies" (IRRI, 1988b) also have traditional counterparts (Biggs, 1980; Goddell, n.d.; King, 1911; RNAM, 1983). Farmers do not adopt mechanical power technologies (e.g., tractors, diesel or electric pumps, axial flow threshers, and mechanical dryers) until they become more profitable to use than traditional alternatives (Pingali et al., 1987).

Scientists have followed farmers in chemical insect pest control. Until 1977, the Philippine government recommended a foliar spray volume of 1000 l/ha for a rice crop that had completed tillering (PCARR, 1977). Rainfed rice farmers applied 135 l/ha (Litsinger et al., 1980). Subsequent research found that all treatments using from 190 to 950 l/ha of perthane equally controlled the brown planthopper, *Nilaparvata lugens* (Aquino and Heinrichs, 1978). As a result, the recommended volume was reduced. Litsinger et al. (1980) also found that farmers applied dosages averaging 0.2 kg active ingredient (ai) per hectare, as opposed to recommended rates of 0.75–1.0 kg ai/ha. Research then determined that control of whorl maggot *(Hydrellia philippina)*, leaffolder *(Cnaphalocrocis medinalis)*, and early-growth stemborer *(Scirophaga incertulas)* was achieved near the farmers' dosage (IRRI, 1983), leading to another change in Philippine recommendations to a dosage of 0.4 kg ai/ha.

Farmers have contributed to chemical weed control measures. Butachlor is a commonly used herbicide for weed control in wet-seeded rice. The manufacturer first recommended 1.0 kg ai/ha at six days after seeding. Farmers reported that butachlor is toxic if applied within six days after seeding and applied lower doses at three days *before* seeding. Testing showed that this practice results in good crop stand, no growth reduction, better and broader-spectrum weed control, and good yield (Mabbayad and Moody, 1984, 1985), and the manufacturer's recommendation was modified. IRRI scientists also observed that farmers in Guimba, Nueva Ecija, pour concentrated herbicides directly into the water at the edges of wet-seeded rice fields (or into irrigation inlets) instead of applying them with a backpack sprayer. Research confirmed that the technique is labor saving and effective for weed control (Mabbayad and Moody, 1987).

Farmers have been active in rice variety development. Two of the most successful upland rices in Sierra Leone and Liberia (ROK3 and LAC23) are of local provenance (Richards, 1985). IR60 was named a variety by the Philippine Seed Board after farmers in Mindanao acquired, tested, and then rapidly disseminated the IRRI selection IR13429-299-2-1-3 undergoing testing. A farmer named Indrasan Singh selected and planted seed that was quickly adopted by many farmers in Uttar Pradesh, India. The "variety" is now widely planted and known as Indrasan. One of the only successful improved upland varieties in the Philippines, UPLRi5, was similarly taken and spread by farmers in Batangas Province where the line was being tested. Such examples are similar to the farmer testing and widespread dissemination in India of Mahsuri (later certified), IR29, and Sarjoo 49 (Maurya, 1989).

Agricultural scientists rarely acknowledge that such technology development is built upon farmer experimentation. There may be several reasons: scientists are under pressure to come up with their own innovations and tend to "forget" farmer origins of technologies; borrowing from farmers is considered "unscientific" (R. E. Rhoades, personal communication); or researchers accept the idea that little can be gained through the study of traditional agriculture because even sustainable traditional systems have not been able to increase production at rates matching increased demand (Ruttan, 1988; Schultz, 1964).

Post–Green Revolution Integration of Formal and Farmer Science

Lack of sustainability of some irrigated rice systems (Pingali et al., 1990), low or declining productivity of marginal areas, and increasing population and demand for rice have made new research approaches necessary. For irrigated and favorable rainfed lowland rice, research has tried to reduce production costs and maintain or increase efficiency of input use. One relatively visible initiative to which farmers have contributed is the development of integrated pest management strategies.

Most irrigated rice farmers in central Luzon who scout their fields for insect pests do so from their field dikes rather than from within the field. Farmers flush stem borer moths from the weedy field borders to determine when to spray (Bandong and Litsinger, 1987). Subsequent research found that border counting of moths serves as an early warning of egg laying in the field (IRRI, 1989a).

Rice research is also shifting to marginal rainfed lowland and upland ecosystems (IRRI, 1989b). As a result, there has been greater appreciation of the fact that "traditional peasant systems of agriculture are not primitive leftovers from the past, but are, on the contrary, systems finely tuned and adapted, both biologically and socially, to counter the pressures of what are often harsh and inimical environments" (Haskell et al., 1981:40). Examples of research that actively tries to build on farmer knowledge and technologies include work on green manures, crop establishment and weed control, and soil erosion control in the uplands.

Green Manures

A recent symposium reconfirmed that green manures can increase rice yields and may contribute to the sustainability of rice-based systems (IRRI, 1988a). Yet there is evidence that "green manures have always loomed larger in the agronomist's mind than in the farmer's, except in parts of China" (Norman, 1982).

In the Philippines, farmers' adoption of green manures as a response to research recommendations is essentially nonexistent. *Indigofera tinctoria*, however, is a traditional green manure used by rainfed lowland rice farmers in northwestern Luzon Island (Bantilan et al., 1989). Historical factors contributed to its adoption. Indigo was introduced and grown during the Spanish colonial period to produce dye, and farmers learned of the plant's fertilizer properties. Although seed scarcity was the main constraint to the use of indigo as a green manure, seed was initially supplied by other farmers who grew indigo as a source of fuel for the tobacco flue-curing industry.

Indigo is compatible with the farmers' cropping systems. After harvests of rice in September and October, or of a postrice garlic or onion crop in February, farmers intercrop (mix seed) indigo with tobacco, mung bean, or maize. Indigo grows slowly during the upland crop period (up to April), survives the dry season, and then produces substantial biomass during the dry-wet transition period in May and June and is incorporated at rice land preparation. Farmers reported that indigo does not compete with the upland crop and adds no additional labor (and may even reduce hand weeding needed in the rice crop). Farmers applied less phosphorus and potassium and used 24–50 kg of inorganic nitrogen per hectare after indigo, as compared with 86–136 kg/ha without. The utilization of indigo resulted in higher returns to cash and equal returns to labor than its nonuse. Seed availability is a constraint because indigo is often incorporated prior to seeding, germination is poor if rains occur at podding, and seed processing is difficult.

Scientific experiments that reflect an understanding of the farmers' system are now being conducted on indigo and other green manure species (Bantilan et al., 1989; Garrity et al., 1989) in the Philippines. Earlier analyses of indigenous green manure systems in China, Taiwan, Bangladesh, the Philippines, and four different regions of India led to both a better-focused research agenda for green manures and a decision tree (Figure 4.1) of minimal criteria for green manure production in particular environments (Garrity and Flinn, 1988).

Research on the use of legumes to increase system productivity and sustainability through crop intensification and more efficient nutrient cycling can be examined at a broader level. Intensive rice-growing systems developed in areas with large populations and controllable water; that is, China, Korea, Japan, Java, and Sri Lanka (Boserup, 1965; Geertz, 1963; Gourou, 1953; Grist, 1959; King, 1911; Leach, 1959; Murphey, 1957; Witfogel, 1957). Legume green manures, mulches, composts, night soil, multiple cropping, intercropping, and relay cropping were widely used in China and Japan at the turn of the century (King, 1911). Farmers appear to develop such systems naturally given the right conditions (e.g., high population, water control, lack of inorganic fertilizers).

Crop Establishment and Weed Control

Researchers observed farmers from Batangas Province in the Philippines using a five-tined wooden harrowlike implement, the *lithao,* to open furrows. Rice was broadcast and harrowed into the furrows (resulting in emergence in rows), and the lithao was used for subsequent interrow cultivation. Agricultural engineers tried to convert the lithao into a furrower and seeder (Aban et al., 1986; Khan et al.,

Figure 4.1. Decision tree: the minimum necessary criteria for green manure production in a particular environment

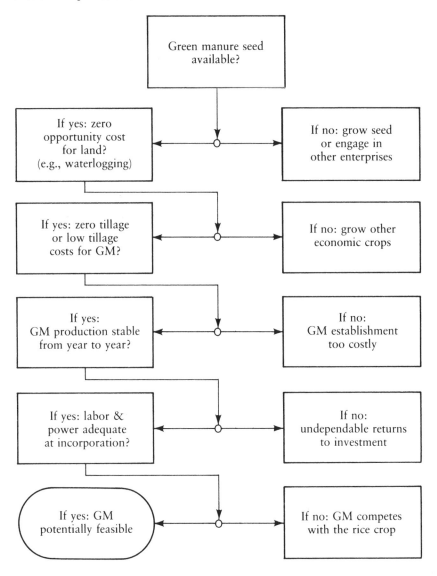

Table 4.2. Rice production labor, Claveria, Misamis Oriental; and Tupi, South Cotabato, Philippines

Activity	Claveria[a]	Tupi[b]
Land preparation	31	16
Seeding	8	2
Interrow cultivation	4	8
Hand weeding	35	19
Harvest and thresh	42	49
TOTAL	120	94

[a] Land is prepared by plow and harrow and then furrowed by plow; seed is drilled; interrow cultivation is by plow.

[b] Land is prepared by plow, harrow, and panudling, then furrowed by panudling; seed is broadcast; and interrow cultivation is by panudling or halodhod.

1971). Work was discontinued because the metal seeder was laterally unstable and tended to bulldoze as weed residues collected around the blades. Agronomists tested the lithao for stand establishment, comparing lithao and broadcasting (IRRI, 1975), and later as a furrow opener under upland and rainfed lowland conditions. The wooden lithao was subject to breakage in everything but sandy Batangas soils, however, and was eliminated from testing.

Other researchers noted the successful use of a farmer-improved steel lithao in heavier soils in Tupi, South Cotabato. Tupi upland rice farming was examined (Fujisaka, in press) and compared with practices of farmers at an on-farm upland rice research site at Claveria, Misamis Oriental. Tupi farmers—using their *panudling* with removable blades and tines—conducted more on-field operations but used 25 percent less labor overall, and 25 percent less man-plus-animal labor, than Claveria farmers who drilled the seed and used a plow for furrowing and cultivation (Table 4.2).

Yields in the two areas in favorable rainfall years reach 3.5–4 metric tons/ha using improved or modern varieties with up to 30 kg N/ha. Although a portion of the higher labor requirement is due to heavier soils in Claveria, it was hypothesized that the Tupi panudling might be an appropriate labor-saving technology for farmers in Claveria. South Cotabato farmers trained a group of Claveria farmers in the use of the implement and in broadcast seeding for wet season 1989. Based on initial experimentation, Claveria farmers modified the panudling by angling the blades forward to achieve better penetration in the heavier soils, and about sixty interested farmers tested the system in wet season 1990. Effectiveness of the technology, farmer-to-farmer technology transfer, farmer adaptation to local conditions, and farmer adoption are being monitored.

Soil Erosion Control

Claveria farmers reported soil erosion and soil nutrient depletion as a major problem (Fujisaka, 1987). Local methods to control soil erosion were limited. Some technical alternatives included farmers' practices from other upland areas and agroforestry techniques being developed by researchers around the world.

Two farmer soil conservation methods were observed in the Philippines. *Gen-gen*—practiced by shifting cultivators in the north—is a technique in which a terrace is formed behind a sweet potato "hedgerow" after crop residues and a covering layer of soil are placed in shallow contour trenches. In later seasons, strips are planted in tiger grass and, in some cases, fruit trees (Barker, 1984). Farmers in Batangas leave unplanted weedy contour strips (made permanent by rocks removed from the alleys) in their upland sweet potato fields. Annual plowing and cropping result in natural terrace formation. Some strips are planted with a few fruit trees and, more recently, fodder grasses. In both cases, the terraces formed behind the grassy strips require little labor for establishment and, according to farmers, are effective in controlling soil erosion.

Researchers' soil conservation alternatives featured simple ways to establish contour lines (the A-frame), contour bunding and ditching, single or double rows of usually nitrogen-fixing trees, and grasses (Celestino, 1984; Lundgren and Nair, 1985). In a project in Cebu, Philippines, farmers established contour lines, constructed bunds and ditches by plowing and shoveling, and planted *Gliricidia sepium* and *Pennisetum purpureum* hedgerows (World Neighbors, n.d.). Under the auspices of the IRRI's upland rice-based systems research in Claveria, farmers who had tried to solve the problem of soil erosion went to Cebu, where they learned the technology from other farmers. Upon their return they established contour hedgerows and began to experiment with and adapt the technology to local conditions. At the same time, a two-pronged research program was initiated.

The first research area includes monitoring farmers' experimentation and technology adaptation and testing farmer-to-farmer training. Some 150 farmers learned about the technology through farmer-to-farmer training, and more than 50 farmers adopted it and are further modifying the system. In three seasons, farmers tried different spacings and combinations of the initial hedgerow species and added new species. They rejected napier grass as too crop-competitive and legume trees as unnecessary. Overall, these farmers are discovering ways to reduce labor requirements (Fujisaka, 1989). Results also indicate that farmers reject use of nitrogen-fixing tree biomass for mulches or incorporation because of the labor required. Many hesitate to invest in labor, especially for bund construction by plowing and shoveling, and instead would prefer simpler grassy strips.

Agronomic studies, the second area of research, initially focused on such topics as establishment and performance of tree legume hedgerows (Mercado et al., 1989) and use of legume tree biomass as fertilizer for rice (Basri et al., 1990). Ongoing studies, however, are responding to (and confirming) farmers' concerns about labor and the need for a slow-growing, single-purpose (bund-stabilization) hedgerow species.

Work to develop soil conservation measures started with researcher-developed agroforestry technologies but is being forced to return to improving traditional systems owing to technical factors (e.g., hedgerow-crop competition) and labor availability. Because farmers opt for systems more like the traditional gen-gen or Batangas grassy strips, farmers and researchers are now making progress toward an improved traditional system rather than the agroforestry ideal of a system of hedgerows of multipurpose trees that provide improved nitrogen cycling.

In the IRRI's collaboration with national rice research programs, diagnostic surveys that pay close attention to farmers' practices and corresponding technical knowledge have become an integral part of the identification of research needs in the rainfed lowlands of Cambodia and Laos, in upland rice areas of Laos and Indonesia, in rice-wheat systems of India and Nepal, and in different rice environments in Madagascar and Bhutan (Fujisaka, 1991). For example, as part of research to improve crop establishment in the rainfed lowlands, a diagnostic survey was conducted on *beusani,* a practice from Nepal, Bangladesh, and India in which farmers plow and plank dry-seeded rice at about thirty days after emergence as a crop establishment and weed management technique. There is still skepticism that a widely applicable improved technology will result, but the basic idea that understanding traditional technologies is an important goal of increasingly limited research resources has been widely accepted.

Conclusions

All future rice research at IRRI will necessarily—albeit at times implicitly—address the issue of sustainability. With the prospect of any additional big breakthroughs in irrigated rice unlikely, and a slow-down in irrigation investments, increasing land and soil degradation, and increasing population density in marginal environments, rice research will face a major battle to maintain the gains of the Green Revolution in the coming century.

"Farmer-back-to-farmer" or "farmer-first" approaches make sense for rice research if farmer-appropriate technologies are to be developed. Work is now being conducted that should make such approaches more systematic and efficient. Rice researchers are increasingly trying to incorporate farmers' perspectives into the identification of research issues and the setting of research priorities. This chapter has argued the importance of incorporating farmers' science, knowledge, and practices into formal agricultural research. At the IRRI, this has occurred through an evolutionary process which first built on farmers' practices during the Green Revolution era and now focuses on farmers' participatory research.

Bibliography

Aban, M. M., E. U. Bautista, E. J. Calilung, and L. C. Kiamco. "Upland Seeder Development at IRRI." In *MAF-IRRI Technology Transfer Workshop*. Los Baños: Philippine Ministry of Agriculture and Food and the International Rice Research Institute, 1986.

Aquino, G. B., and E. A. Heinrichs. "Spray Volumes in Relation to Brown Planthopper Control." *International Rice Research Newsletter*, 3:21, 1978.

Ashby, J. A. "The Effects of Different Types of Farmer Participation on the Management of On-Farm Trials." *Agricultural Administration and Extension*, 24:234–52, 1987.

Bandong, J., and J. Litsinger. "Development of Action Control Thresholds for Major Rice Pests." In P. Teng and K. L. Heong, eds., *Pesticide Management and Integrated Pest Management in SE Asia*. College Park, Md.: Consortium for International Crop Protection, 1987.

Bantilan, R. T., C. C. Bantilan, and D. P. Garrity. "*Indigofera tinctoria* as a Green Manure Crop in Rainfed Rice-based Systems." IRRI unpublished seminar paper, Los Baños, 1989.

Barker, T. C. "Shifting Cultivation among the Ikalahans." Program on Environmental Science and Management, Working Paper 1. Los Baños, Philippines: University of the Philippines at Los Banos, 1984.

Basri, I. H., A. R. Mercado, and D. P. Garrity. "Upland Rice Cultivation Using Leguminous Tree Hedgerows on Strongly Acid Soils." IRRI, unpublished Saturday Seminar paper, Los Baños, 1990.

Biggs, S. D. "Research by Farmers: The Importance of Informal Agricultural R and D Systems in Developing Countries." Sussex, England: Institute of Development Studies, 1980.

———. "Resource-Poor Farmer Participation in Research: A Synthesis from Nine National Agricultural Research Programs." The Hague: International Service for National Agricultural Research, 1988.

Bonifacio, E., G. Quick, and P. Pingali. "Deriving Design Specifications for Mobile Dryers: An End-User Perspective." Paper presented at the International Agricultural Engineering Conference, Bangkok, Asian Institute of Technology, 1990.

Boserup, E. *Conditions of Agricultural Growth*. Chicago: Aldine Press, 1965.

Box, L. "Experimenting Cultivators: A Methodology for Adaptive Agricultural Research." Agricultural Administration (Research and Extension) Network Discussion Paper 23. London: Overseas Development Institute, 1987.

Brokensha, D., D. M. Warren, and O. Werner, eds. *Indigenous Knowledge Systems and Development*. Washington, D.C.: University Press of America, 1980.

Celestino, A. F. "Establishment of Ipil-ipil Hedgerows for Soil Erosion and Degradation Control in Hilly Land." FSSRI/UPLB-CA Paper. Los Baños: University of the Philippines at Los Baños, 1984.

Chambers, R., A. Pacey, and L. A. Thrupp, eds. *Farmer First: Farmer Innovation and Agricultural Research*. London: Intermediate Technology Publications, 1989.

Farrington, J., and A. Martin. "Farmer Participatory Research: A Review of Concepts and Practices." Agricultural Administration (Research and Extension) Network Discussion Paper 19. London: ODI, 1987.

Fujisaka, S. "Upland Rice Farmers' Soil Nutrient Management: Practice and Technical Knowledge in Claveria, Misamis Oriental." Unpublished paper, dated 1987.

———. "A Method for Farmer-participatory Research and Technology Transfer: Upland Soil Conservation in the Philippines." *Experimental Agriculture*, 25:423–33, 1989.

———. 1991. "A Set of Farmer-based Diagnostic Methods for Setting Post 'Green-Revolution' Rice Research Priorities." *Agricultural Systems*, 36:191–206.

———. "Taking Farmer Knowledge Seriously: Weeding and Seeding Upland Rice in the Philippines." In D. M. Warren, D. Brokensha, and L. J. Slikkeveer, eds., *Indigenous Knowledge Systems: The Cultural Dimension of Development*. London: Kegan Paul International, in press.

Garrity, D. P., R. T. Bantilan, C. C. Bantilan, P. Tin, and R. Mann. "*Indigofera tinctoria*: A Farmer-proven Green Manure for Rainfed Ricelands." Paper presented at the Bio- and Organic Fertilizers Symposium, University of the Philippines at Los Baños, October 9–12, 1989.

Garrity, D. P., and J. C. Flinn. "Farm Level Management Systems for Green Manure Crops in Asian Rice Environments." In *Green Manure in Rice Farming*. Los Baños, Philippines: IRRI, 1988.

Geertz, C. *Agricultural Involution: The Process of Ecological Change in Indonesia*. Los Angeles: University of California Press, 1963.

Goodell, G. n.d. "Communication from Farmer to Researcher." Unpublished paper, IRRI.

Gourou, P. *The Tropical World*. New York: Longmans Green, 1953.

Grist, D. H. *Rice*. London: Longmans Green, 1959.

Haskell, P. T., T. Beacock, and P. J. Wortley. "World-wide Socio-economic Constraints to Crop Protection." In T. Kommedahl, ed., *Proceedings of the Ninth International Congress on Plant Protection*, vol. 1, pp. 463–66. Minneapolis, Minn.: Burgess Publishing.

——. "Annual Report for 1974." Los Baños, Philippines: IRRI, 1975.

——. "Insecticide Evaluation for 1977." Los Baños, Philippines: IRRI Entomology Department, 1977.

——. "Insecticide Evaluation for 1982." Los Baños, Philippines: IRRI Entomology Department, 1983.

——. "Insecticide Evaluation for 1986." Los Baños, Philippines: IRRI Entomology Department, 1986.

——. "Farm and Post-harvest Machinery Developments: Semi-annual Progress Report no. 87–2. Los Baños, Philippines: IRRI Agricultural Engineering Department, 1987.

——. *Green Manure in Rice Farming*. Los Baños, Philippines: IRRI, 1988a.

——. "Farm and Post-harvest Machinery Development: Semi-annual Progress Report no. 88-1. Los Baños, Philippines: IRRI Agricultural Engineering Department, 1988b.

——. *IRRI Research Highlights*. Los Baños, Philippines: IRRI, 1989a.

——. *IRRI: Toward 2000 and Beyond*. Manila: International Rice Institute, 1989b.

Khan, A. U., F. E. Nichols, and B. Duff. "Semi-annual Progress Report No. 12." Agricultural Engineering Department, IRRI, 1971. Typescript.

King, F. H. *Farmers of Forty Centuries*. Emmaus, Pa.: Rodale Press, 1911.

Leach, E. "Hydraulic Society in Ceylon." *Past and Present*, 15:2–25, 1959.

Litsinger, J. A., E. C. Price, and R. T. Herrera. "Small Farmer Pest Control Practices for Rainfed Rice, Corn, and Grain Legumes in Three Philippine Provinces." *Philippine Entomologist*, 4(1–2):65–86, 1980.

Lundgren, B., and P. K. R. Nair. "Agroforestry for Soil Conservation." In S. A. El Swaify, W. C. Moldenhauer, and A. Lo, eds., *Soil Erosion and Conservation*. Ankeny, Iowa: Soil Conservation Society of America, 1985.

Mabbayad, M. O., and K. Moody. "Effect of Time of Herbicide Application on Crop Damage and Weed Control in Wet-Seeded Rice." *IRRI Rice Research Newsletter*, 9(3):22, 1984.

——. "Improving Butachlor Selectivity and Weed Control in Wet-Seeded Rice." *Journal of Plant Protection in the Tropics*, 2(2):117–24, 1985.

——. "Using Undiluted Herbicides for Weed Control in Wet-Seeded Rice." Paper presented at the International Conference on Pesticides in Tropical Agriculture, Kuala Lumpur, Malaysia, 1987.

Matlon, P., R. Cantrell, D. King, and M. Benoit-Cattin, eds. *Farmers' Participation in the Development of Technology: Coming Full Circle.* Ottawa: International Development Research Center, 1984.

Maurya, D. M. "The Innovative Approach of Indian Farmers." In R. Chambers, A. Pacey, and L. A. Thrupp, eds., *Farmer First.* London: Intermediate Technology Publications, 1989.

Mercado, A. R., A. M. Tumacas, and D. P. Garrity. "The Establishment and Performance of Tree Legume Hedgerows in Farmers' Fields in a Sloping Upland Environment." Paper presented at the Fifth Annual Meeting of the Federation of Crop Science Societies of the Philippines, 1989.

Murphey, R. "The Ruin of Ancient Ceylon." *Journal of Asian Studies,* 16:181–200, 1957.

Norman, M. J. T. "A Role for Legumes in Tropical Agriculture." In P. H. Graham and S. C. Harris, eds., *Biological Fixation in Tropical Agriculture.* Cali, Colombia: Centro Internacional de Agricultura Tropical, 1982.

Philippine Council for Agricultural and Resources Research (PCARR). *The Philippines Recommends for Rice.* Los Baños, Philippines: PCARR, 1977.

Pingali, P. L., Y. Bigot, and H. P. Binswanger. *Agricultural Mechanization and the Evolution of Farming Systems in Sub-Saharan Africa.* Baltimore and London: Johns Hopkins University Press, 1987.

Pingali, P. L., P. F. Moya, and L. E. Velasco. "The Post Green Revolution Blues in Asian Rice Production: The Diminished Gap between Experiment Station and Farmer Yields." IRRI Social Science Division Paper 90-01, 1990.

Regional Network for Agricultural Machinery (RNAM). *Testing, Evaluation, and Modification of Weeders.* Manila: RNAM/ESCAP, 1983.

Rhoades, R. E. *Breaking New Ground: Agricultural Anthropology.* Lima: International Potato Center, 1984.

——. "Farmers and Experimentation." Agricultural Administration (Research and Extension) Network Discussion Paper 21. London: Overseas Development Institute, 1987.

Rhoades, R. E. and R. Booth. "Farmer-Back-to-Farmer: A Model for Generating Acceptable Agricultural Technology." *Agricultural Administration,* 11:127–37, 1982.

Richards, R. *Indigenous Agricultural Revolution.* London: Hutchinson, 1985.

Ruttan, V. "Sustainability Is Not Enough." *American Journal of Alternative Agriculture,* 3:128–30, 1988.

Sagar, D., and J. Farrington. "Participatory Approaches to Technology Generation: From the Development of Methodology to Wider-Scale Implementation." Agricultural Administration (Research and Extension) Network Paper 2. London: Overseas Development Institute, 1988.

Schultz, T. *Transforming Traditional Agriculture.* New Haven: Yale University Press, 1964.

Thurston, H. D. "Plant Disease Management Practices of Traditional Farmers." *Plant Disease,* 74(2)96–102, 1990.

Warren, D. M., and K. Cashman. "Indigenous Knowledge for Sustainable Agricultural Development." Gatekeeper Series SA10. London: International Institute for Environment and Development, 1988.

Wittfogel, K. *Oriental Despotism*. New Haven: Yale University Press, 1957.

World Neighbors. *Simple Soil and Water Conservation Methods for Upland Farms*. Cebu City, Philippines: World Neighbors, n.d.

5

Constraints on Nitrogen Fertilizer Use on Sorghum in Semiarid Tropical India: Rainy-Season Hybrids

Karen Ann Dvořák

Introduction

Fertilizer use on dryland crops in the semiarid tropics (SAT) of India remains very low in spite of good agronomic responses to fertilizer in on-station experiments and multilocational trials. In 1985, the International Fertilizer Development Center (IFDC), in collaboration with the International Crops Research Institute for the Semi-Arid Tropics (ICRISAT), initiated a project to investigate on-farm constraints on the use of nitrogenous fertilizer on sorghum *(Sorghum bicolor)* in the SAT. This chapter covers fertilizer use on rainy-season (June–October) sorghum hybrids in the Vidharbha region of Maharashtra.

In farming systems research, survey methods have become associ-

This chapter is an expanded version of my article "On-Farm Experiments as a Diagnostic Method: Constraints to Nitrogen Fertilizer Use on Sorghum in Semi-arid Tropical India," *Experimental Agriculture*, 28, 1992. Ulrike Piepenbrink, V. K. Chopde, H. Nawaz, Auke deBoer, and T. S. Walker all made invaluable contributions to the fieldwork. Shri C. N. Alankare and D. G. Holey were very helpful with necessary arrangements for accommodations in the nearby market town of Murtizapur. An abbreviated description of the research methods was part of a symposium presentation, "Setting Research Priorities for the IARCs: A New Role for the Agricultural Economist," American Agricultural Economics Association Annual Meetings, August 2–5, 1987, East Lansing, Michigan.

ated with the diagnostic stage of the research process, and on-farm trials with the technology-testing and adaptation stages. Our approach differed. On-farm, diagnostic fertilizer experiments and topical surveys drawing on farmers' knowledge and experience were designed to complement existing data and examine the use of fertilizer in the farmers' context. A detailed data set on farmer production practices was available (Binswanger and Jodha, 1978; Singh et al., 1987), as were results from fertilizer experiments conducted by IFDC/ICRISAT at ICRISAT Center (Buresh et al., 1984; Hong, 1986; Moraghan et al., 1984a, 1984b). In addition, results from numerous national fertilizer trials conducted at various locations and times were on record.

The Production Environment

The Vidharbha region of India is characterized by medium-deep Vertisols and annual rainfall of 800–1200 mm. The rainfall pattern is less erratic than elsewhere in the SAT, and the region is considered to have high potential for sorghum. Over the course of the past decade, adoption of sorghum hybrids and improved varieties has increased in this region. In several districts the area planted in sorghum hybrids exceeds 50 percent of the total sorghum area, and Jansen (1988) estimated that in thirteen districts, adoption ceilings will eventually exceed 80 percent. Nevertheless, even in districts with more than half of the sorghum area under hybrids, average yields are 0.8–1.5 metric tons/ha (Government of India, n.d.), compared with research station potentials of 6 metric tons/ha.

Relatively low levels of fertilizer use account for at least part of the difference. Current recommendations for the Akola subregion are 80 kg N/ha and 40 kg P_2O_5/ha (ICAR/AICRPDA, 1983:15). For the Indore subregion, recommendations are 50 kg N/ha and 25 kg P_2O_5/ha (ibid.:28). Time-series data on fertilizer use by crop are not readily available, but special studies indicate that fertilizer use is well below recommended levels. Averages as low as 18 kg N/ha on fertilized fields have been reported (Jha and Sarin, 1984). Data for Akola district for 1970–77 showed no significant upward trend in nitrogen fertilizer use (Jha et al., 1981). Nevertheless, fertilizer use has been reported to be higher on hybrid than on local sorghum for the same area (Jha and Sarin, 1984).

Kanzara village, located in Akola district, was selected as a study village. About 40 percent of the area devoted to sorghum is under hybrids. An early study of village data found that, averaging data for 1975–78, nitrogen was applied to 46 percent of the hybrid sorghum area; the average nitrogen application rate for fertilized hybrid sorghum was 28.5 kg/ha (Jha and Sarin, 1984). Between 1975–76 and 1985–86, fertilizer use increased from less than 10 to 30 kg N/ha (averaged over all crops, irrigated and dryland).

It has been argued that fertilizer scarcity constrains its use in some parts of India (Desai, 1986). Of the three long-term ICRISAT study villages, Kanzara was deemed most likely to address the supply constraint hypothesis because fertilizer is used regularly there. Although no formal research had been done on this issue, observers and enumerators living in the village reported complaints from farmers about fertilizer availability.

A second hypothesis was that because village soils are variable, recommendations for optimal use based on uniform and relatively good quality soils are not applicable. According to a standard soil classification system, 85 percent of the arable land in Kanzara is "medium black." Early in 1986, a special study on indigenous soil classification was conducted (Dvořák, 1989). Sixty fields were visited and the farmers were asked to give the soil names and characteristics. These data were organized into a village soils taxonomy with three major and four secondary soil types. The three major soil types—Bhari kali, Madhyam kali, and Halki—covered 17 percent, 31 percent, and 13 percent of the area, respectively.

Methods

A meeting was held with Kanzara farmers prior to the 1986 growing season. The researchers explained their interest in doing experiments on unirrigated hybrid sorghum and solicited suggestions on types of experiments. Because Kanzara farmers had experience with fertilizer, the type of experiment was left open to discussion. We stipulated only that research should deal with unirrigated hybrid sorghum. Farmers suggested experiments on fertilizer rates, timing of top-dressing, and comparison of urea with complex fertilizers. The first two topics were selected because of their relative simplicity and the degree of interest expressed. Before the experiments were planted, soil samples and data on soil type, field characteristics, and manuring history were taken. Fertilizer for the dosage experiments was provided, but

Table 5.1. Plot characteristics for dosage experiments, Kanzara village, Maharashtra, India, rainy season, 1986

Plot	Farmer soil group	Soil depth[a] (cm)	N[b] (μg)	Organic carbon[c] (%)	Years elapsed since most recent application of farmyard manure
83A	Halki	15	14.6	0.51	2
55F	Halki	22	16.1	0.29	2
54C	Madhyam	45	26.4	0.58	1
48A	Madhyam	30	8.4	0.29	1
32A	Madhyam	45	14.3	0.46	1
85A	Madhyam	30–45	25.7	0.67	0
57B	Bhari	140	11.6	0.56	3

[a] Specified by farmer.
[b] Mineralizeable nitrogen as measured by accumulation of ammonium after 7 days of anaerobic incubation at 40° C (Waring and Bremner, 1964; Keeney, 1982: 727–28).
[c] Dichromate oxidation without external heat (Walkley-Black).

all other expenses and risks were borne by the farmers, who managed the entire fields according to their usual preferred practices. Researchers supervised harvesting of the treatments, drying and threshing the grain, and weighing the grain and fodder in toto for each subplot.

Dosage experiments were conducted in seven fields. Subplots with five levels of nitrogen—0, 28, 57, 85, and 114 kg N/ha—were replicated twice in 10-m × 10-m subplots by basal application of urea in seed rows. (The nitrogen levels correspond to measures used by farmers; i.e., 0.5, 1, 1.5, and 2 bags urea per acre.) Plots offered were from the three major soil types with a range of manuring histories (Table 5.1).

Three farmers participated in an experiment on timing of top-dressing. The fertilizer 20:20:0 was applied basally in seed rows at 124 kg/ha (25 kg N/ha) to all subplots. The three treatments were control (no subsequent fertilizer application) and application of 57 kg N/ha as urea at thirty (recommended) and forty-five (traditional) days after planting (DAP).

Results

Rainfall during the 1986 rainy season was 550 mm, well below the average of 817 mm/acre. Response to nitrogen fertilizer in the dosage

Table 5.2. Estimated response of sorghum (SPH 221) grain yield (Y) to nitrogen fertilizer application (N) and ex post optimal N level (selected farmers' fields, Kanzara village, Akola district, Maharashtra, India, rainy season, 1986)

Plot	Soil group	Estimated function Y (kg grain/ha) N (kg N/ha)	r^2	Ex post optimal N level (kg N/ha)
83A	Halki	$Y = 235 + 39.887 \ln N$	0.46	11
55F	Halki	$Y = 45 + 6.203N - 0.051N^2$	0.65	26
54C	Madhyam	$Y = 1928 + 117.05 \ln N$	0.37	33
48A	Madhyam	$Y = 478 + 18.834N - 0.131N^2$	0.66	58
32A	Madhyam	$Y = 271 + 9.01N - 0.042N^2$	0.81	66
95A	Madhyam[a]	$Y = 2633 - 2.696N$	0.05	0
57B	Bhari	$Y = 713 + 15.288N - 0.052N^2$	0.89	113

Note: Optimal N level based on grain price of Rs 1.3/kg and N price of Rs 456 kg N/ha.

[a] Freshly manured.

trials varied widely between fields in yield, functional form, and rate of response, illustrating the (in)famous "soil heterogeneity" of the SAT (Table 5.2). Average control yields ranged from 0.4 to 25.7 kg/(100 m²). Quadratic, log-linear, and linear forms of response occurred.[1]

Ex post optimal fertilizer rates were calculated based on previous nitrogen and grain sorghum prices. The sorghum price at harvest in the nearby market town of Murtizapur was Rs 130/100 kg (quintal), or Rs 1.3/kg. The price of urea was Rs 210/kg, or Rs 4.56/kg N.[2] The optimal fertilizer application rate ranged from 0 to 113 kg N/ha. With the exception of one plot, optimal fertilizer levels were consistently grouped by farmer soil type. For the plot with the ideal soil type, Bhari kali, ex post optimal fertilizer level was the highest at 113 kg N/ha. Optimal fertilizer level for the Madhyam ("medium") kali soils fell in between, at 33–66 kg N/ha. Neither organic carbon, mineralizeable nitrogen, nor a simple index of soil depth is as satisfactory for stratification.[3] The farmer soil-classification system includes soil depth,

[1] Any of these forms is justified by theory, depending on the range of nitrogen values tested and the field conditions. The choice of functional form was made by examining data plotted graphically for each field.

[2] Marketing costs have not been added because all farmers visit Murtizapur on Fridays for marketing activities. It would be inappropriate to attribute these costs specifically to the purchase of fertilizer or marketing of sorghum. The addition of a portion of the costs to fertilizer and sorghum prices could affect the conclusions regarding the absolute optimum fertilizer level but would not affect the conclusion that there exist different optima for different soil groups.

[3] A "phosphorus index" would not be expected to be a powerful indicator in these soils. Nevertheless, available phosphorus was tested. The nature of the nitrogen response was not correlated with available phosphorus.

Table 5.3. Average use of nitrogen fertilizer (kg N/ha) on unirrigated hybrid sorghum, Kanzara village, Maharashtra, India, 1975–76 through 1984–85

Year	1975–76	1976–77	1977–78	1978–79	1979–80	1980–81	1981–82	1982–83	1983–84	1984–85
N	25	19	20	24	32	34	33	36	44	47

but other characteristics are also important; for example, color, fertility, and moisture-holding capacity. The data support the hypothesis that farmer soil types stratify fertilizer recommendation domains.

Weighing the calculated optima by percentage distribution of area for Bhari kali, Madhyam, and Halki soils results in a village average of 50–70 kg N/ha. This is somewhat high because the excluded soil groups are more closely related to Madhyam kali and Halki than to Bhari. For an across-year ex post optimum, it is low because rainfall in the year of the experiment was below average.

The exception to the stratification by farmer soil type occurred only for the freshly manured plot (85A). Manure had been applied at a rate of 4500 kg/ha, roughly 90 g N/ha. Control yield was 2.5 metric tons/ha, and there was no significant response to nitrogen fertilizer.

In the timing experiment, results were mixed. Top-dressing improved yields for the two fields with control (basal application of fertilizer only) yields under 2 metric tons/ha. For one plot, response to top-dressing was poorer when the application date was advanced. For the second plot, there was no significant difference in response to the change in top-dressing date.

Discussion

For all but the field with the best quality soil, the ex post optimal level of nitrogen fell below the recommended level of 80 kg N/ha. Data on fertilizer use in Kanzara (Table 5.3) show that farmers apply less than the optimal level of nitrogen. Farmers complained about fertilizer availability, yet such complaints were registered even when urea was available in the local market. Farmers distinguished between urea and complex fertilizers, felt that complex fertilizers must be "better," and so reported "fertilizer shortages" when complex fertilizers were not available at the fixed government price. The preference for complex fertilizers could be associated with the use of complex fertilizers on cotton, the major cash crop in the region.

Somewhat surprisingly, the farmers interviewed were not conversant with the concept of elemental composition of fertilizers; they calculated application strictly on the basis of bags of fertilizer per acre. At current levels of fertilizer use, nitrogen response could be obtained by using the equivalent in urea on sorghum, but farmers were unaware of this possibility. Sorghum hybrids are relatively new, and some farmers in the village were on their third generation of hybrid in ten years, insufficient time to have evaluated the response of individual cultivars to fertilizer use in a variety of rainfall years.

Response to nitrogen fertilizer was absent only in the freshly manured plot. The six remaining plots had been manured at least once in the preceding three years, but there was no consistent correlation between the manuring history and yield or nitrogen response in the experiment, suggesting that manuring affects the response of sorghum to nitrogen fertilizer application only in the first season after manure application. This is consistent with the observation that the quality of farmyard manure is poor and with the hypothesis that nitrogen loss in storage and decomposition is rapid during the first rainy season after application. The data are not conclusive, but they point the way for future research on fertility management using organic materials and fertilizer.

The participating farmers had expressed an interest in testing an earlier top-dressing date because fertilizer application, particularly when bullocks are used, is easier when plants are shorter. Data are available for only three fields but may be examined in conjunction with the dosage and on-station experiments. The lack of response to top-dressed fertilizer in the field with a control yield greater than 2 metric tons/ha is consistent with the response patterns in the dosage experiment. Of the remaining two fields in the top-dressing experiment, the efficiency of top-dressed fertilizer was greater for the shallower soil. The reverse was true for the basal fertilizer applied in the dosage experiments; that is, efficiencies for the shallower soils were markedly lower than those for the deeper soils. In experiments conducted over a number of years at ICRISAT Center, very high technical fertilizer efficiencies were obtained across a range of soil depths and rainfall years, with one exception: fertilizer efficiency declined notably in shallow soils in high-rainfall years (Katyal et al., 1987).

As noted, 1986 was low in rainfall. Because of the apparent inconsistencies, distribution of rainfall was examined using weekly average rainfall probabilities for the nearby Akola meteorology station (Virmani et al., 1982:18–19). Probable rainfall decreases at standard week

Figure 5.1. Mean weekly rainfall, Akola station, Maharashtra, India.

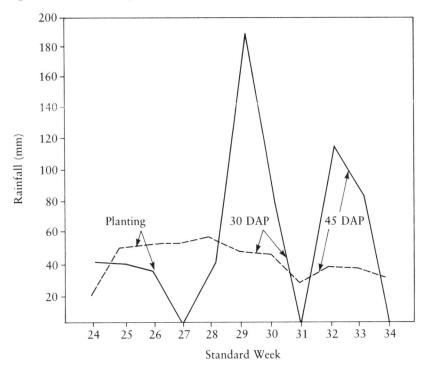

Based on thirty years' rainfall data (Virmani et al., 1982, based on data from the Indian Meteorological Department, Poona, India) and weekly rainfall in the rainy season, 1986, Kanzara village, Akola district, Maharashtra, India. DAP = days after planting.

30 (Figure 5.1). Using the ICRISAT production records, it was determined that planting usually takes place in standard week 24, in which case traditional timing of top-dressing falls in standard week 21, or after the drop in rainfall. Advancing the time of top-dressing therefore increases the probability that intense rainfall will occur within a short time thereafter.

The low efficiencies on shallow soils reported for high-rainfall years may actually be associated with intense rainfall within a short time after fertilizer application (basal and top-dressed). Daily rainfall records for 1986 show that early season rainfall was favorable, but no intense storms occurred until nineteen days after planting. Intense rainfall occurred again at nine days after early top-dressing. In contrast, rainfall was quite steady in the week following late top-dressing.

This could explain the higher efficiency of the shallower soil in the diagnostic experiment. In most years, the traditional time of top-dressing—forty-five days after planting—is more likely to avoid a high rainfall period and concomitant decline in fertilizer efficiency.

Overall, fertilizer efficiencies were more sensitive to rainfall distribution on shallow soils. Thus, designation of a year as "high" or "low" rainfall is insufficient to interpret the results of the dosage and top-dressing experiments. More specific information on fertilizer efficiency in relation to the time lag between fertilizer application and intense rainfall is required.

Conclusions

Farmer-designated soil groups provided a useful basis for interpreting fertilizer responses across plots with a high degree of heterogeneity, and for defining fertilizer recommendation domains. Soil variability explains, to some extent, why farmers use less than the recommended amounts of nitrogen fertilizer for hybrid sorghum. In place of the current blanket recommendation to farmers, recommendations for both deep and shallow soils would facilitate farmers' efforts to adjust fertilizer application for consistently favorable returns.

Fertilizer "shortages" play a role in below-expected levels of fertilizer use, but with a twist. Because of successive releases of hybrids and the variability in rainfall from year to year, ten years appears to have been insufficient for farmers to fully appreciate the nitrogen responsiveness of sorghum hybrids. Farmers' preferences for complex fertilizers, coupled with a lack of familiarity with fertilizer nutrient contents, have also contributed to suboptimal use of nitrogen. Moreover, this preference leads to reports of "fertilizer shortages" when complex fertilizers are unavailable, even though urea is available.

This "constraint" on fertilizer use may be expected to resolve itself over time, as farmers gain more experience using fertilizer on sorghum hybrids in different rainfall conditions. The current government policy of allocating urea to this region could be expected to act as an incentive to farmers to increase their use of nitrogen in relation to phosphorus and potassium. Extension or marketing programs on elemental fertilizer composition would be a positive incentive in the same direction.

Evidence bearing on differences in crop uptake of fertilizer nitrogen for different soil depths was not conclusive. Additional research on the effect of rainfall patterns—frequency and intensity—in the first sixty days after planting is needed to develop recommendations on method of application (basal vs. split) and timing of top-dressing for the shallower soils that constitute a significant portion of the sorghum production area. Basic research on the relative importance of denitrification and leaching as loss mechanisms would complement applied research on application methods and timing.

The on-farm experiments were a crucial component in diagnosing the constraints on fertilizer use in Kanzara. In-field replication and a limited number of experimental fields permitted frequent, detailed researcher observations and increased confidence in experimental results. The experimental results provided the lines of inquiry on fertilizer response in the farmers' environment, which were pursued using base data, on-station trial results, and topical surveys.

The surveys on fertilizer use and soil type, supplemented by frequent informal discussions with farmers, were also crucial in interpreting the experimental data. The time-series data on fertilizer use indicate that farmers are gradually increasing their use of nitrogen fertilizer as they observe the response of sorghum hybrids to fertilizer. The farmers' awareness of soil depth as an important soil descriptor supports on-station research on differential fertilizer efficiency by depth. The farmers' knowledge of soil taxonomy provides a reasonable basis for designing and making fertilizer recommendations.

In contrast, the farmers' understanding of the elemental composition of complex fertilizers is poor and represents an area where educational programs could be usefully targeted. Awareness of this weak point in farmer comprehension was important in interpreting their complaints about fertilizer shortages. Taken together, these perceptions of farmers contributed significantly to achieving the diagnostic objective.

Bibliography

Binswanger, H. A., and N. S. Jodha. *Manual of Instructions for Economic Investigators in ICRISAT's Village Level Studies.* Patancheru, Andhra Pradesh, India: International Crops Research Institute for the Semi-Arid Tropics, 1978.

Buresh, R. J., P. L. G. Vlek, and J. M. Stumpe. "Labeled Nitrogen Fertilizer Research with Urea in the Semi-arid Tropics. I: Greenhouse Studies. *Plant and Soil,* 80:3–19, 1984.

Desai, Gunvant M. "Policies for Growth in Fertiliser Consumption: The Next Stage." *Economic and Political Weekly,* 21:928–33, 1986.

Dvořák, Karen Ann. "Indigenous Soil Classification in Semi-arid Tropical India." Economics Group Progress Report 84. Patancheru P.O., Andhra Pradesh, India: International Crops Research Institute for the Semi-Arid Tropics, 1989.

———. "On-Farm Experiments as a Diagnostic Method: Constraints to Nitrogen Fertilizer Use on Sorghum in Semi-arid Tropical India." *Experimental Agriculture,* 28, 1992.

Government of India. *Agricultural Situation in India.* New Delhi: Directorate of Economics and Statistics, Ministry of Agriculture, various years.

Hong, C. W. *Work Report. IFDC/ICRISAT Cooperative Research on the Fate and Efficiency of Fertilizer Nitrogen in the Cropping Systems of Semiarid Tropics—India.* Muscle Shoals, Ala.: International Fertilizer Development Center, 1986.

ICAR/AICRPDA (Indian Council of Agricultural Research/All India Coordinated Research Project for Dryland Agriculture). *Improved Agronomic Practices for Dryland Crops in India.* 3d ed. Hyderabad, India: AICRPDA, 1983.

Jansen, Hans. "Adoption of Modern Cereal Cultivars in India: Determinants and Implications of Interregional Variation in the Speed and Ceiling of Diffusion." Ph.D. diss., Cornell University, Department of Agricultural Economics, 1988.

Jha, D., S. K. Raheja, R. Sarin, and P. C. Mehrotra. "Fertilizer Use in Semi-arid Tropical India: The Case of High Yield Varieties of Sorghum and Pearl Millet." ICRISAT Economics Program Progress Report 22. Patancheru, Andhra Pradesh, India: International Crops Research Institute for the Semi-Arid Tropics, 1981.

Jha, D., and R. Sarin. *Fertilizer Use in Semi-arid Tropical India.* ICRISAT Research Bulletin 9. Patancheru, Andhra Pradesh, India: International Crops Research Institute for the Semi-Arid Tropics, 1984.

Katyal, J. C., C. W. Hong, and P. L. G. Vlek. "Fertilizer Management in Vertisols." Paper presented at the First Regional Seminar on Management of Vertisols under Semi-arid Conditions, December 1–6, 1986, Nairobi, Kenya. Muscle Shoals, Ala.: International Fertilizer Development Center, 1987.

Keeney, D. R. "Nitrogen Availability Indices." In A. L. Page, ed., *Soil Analysis Part 2. Chemical and Microbiological Properties.* Agronomy Monograph no. 9, 2d ed. Madison, Wis.: American Society of Agronomy, and Soil Science Society of America, 1982.

Moraghan, J. T., T. J. Rego, R. J. Buresh, P. L. G. Vlek, J. R. Burford, S. Singh, and K. L. Sahrawat. "Labeled Nitrogen Fertilizer Research with Urea in the Semi-arid Tropics. II. Field Studies on a Vertisol." *Plant and Soil,* 80:21–33, 1984a.

Moraghan, J. T., T. J. Rego, and R. J. Buresh. "Labeled Nitrogen Fertilizer Research with Urea in the Semi-arid Tropics. II. Field Studies on an Alfisol." *Plant and Soil,* 82:193–203, 1984b.

Singh, R. P., H. P. Binswanger, and H. S. Jodha. *Manual of Instructions for Economic Investigators in ICRISAT's Village Level Studies.* Patancheru, Andhra Pradesh, India: International Crops Research Institute for the Semi-Arid Tropics, 1987. Revised.

Virmani, S. M., M. V. K. Siva Kuma, and S. J. Reddy. *Rainfall Probability Esti-*

mates for Selected Locations of Semi-arid India. Research Bulletin no. 1, 2d ed. Patancheru, Andhra Pradesh, India: International Crops Research Institute for the Semi-Arid Tropics, 1982.

Virmani, S. M., R. M. Willey, and M. S. Reddy. "Problems, Prospects, and Technology for Increasing Cereal and Pulse Production from Deep Black Soils." In *Improving the Management of India's Black Soils for Increased Production of Cereals, Pulses, and Oilseeds.* New Delhi, India: Patancheru, Andhra Pradesh, India: International Crops Research Institute for the Semi-Arid Tropics, 1981.

Waring, S. A., and J. M. Brenner. "Ammonium Production in Soil under Water-logged Conditions as an Index of Nitrogen Availability." *Nature* (London) 201:951–52.

6

Farmer Participation and the Development of Bean Varieties in Rwanda

Louise Sperling

Introduction

The problem is well known and amazingly long-lived: high-yielding, disease-resistant crop varieties, refined and nurtured by discriminating breeders, are rejected by farmers soon after release. The products of years of on-station research are swiftly relegated to the family cooking pot—no storage, no reseeding. Bean varietal development in Rwanda is no exception to this trend. Out of the fifty-odd varieties tested by the Institut des Sciences Agronomiques du Rwanda (ISAR) over the last ten years, perhaps five remain widely grown in farmers' fields. This rate can and should be improved.

Researchers at Third World agricultural institutes are increasingly documenting how farmers' criteria may diverge from breeders' when it comes to selecting varieties. For example, work by the International Potato Center (CIP) in Rwanda identified farmers' preference for short-cycle and often short-dormancy potato cultivars, helping to shift a national breeding program that had been largely screening for later-maturing, long-dormancy varieties (Haugerud, 1987).

For intellectual support and active participation, the author is especially indebted to D. Cishahayo, J. Davis, W. Graf, A. Munyemana, B. Ntambovura, P. Nyabyenda, P. Trutmann, and a group of gifted interviewers in ISAR's Crop Department. J. Moock, R. Rhoades, and R. Herdt offered insightful comments on various drafts. M. Loevinsohn's aid has been more than generous at all stages of research and writing.

Further, farmers themselves may differ in their criteria according to economic means, market orientation, gender, individual preference, and the like. Research by the Centro Internacional de Agricultural Tropical (CIAT), in collaboration with the Instituto Colombiano Agropecuario (ICA), showed that women in Cauca, Colombia, select beans with small grains and high yields—for consumption purposes—while men aim for large-grain, red-colored types that fetch a favorable price (CIAT, 1988). Breeders' strategies, consciously or not, target specific populations. Given the diversity of farmers' preferences and constraints, a tacit decision is made as to whose problems should be given priority.

This chapter focuses on the link—too often a gap—between farmers' and breeders' perspectives in varietal development, addressing the question of how the two might be melded. In Rwanda (Voss and Graf, 1991) as elsewhere (for example, CIAT/Colombia; see Ashby et al., 1987), scientists have been using on-farm varietal evaluations as one means of feeding information from rural fields to research stations. As the last stage in varietal selection, farmers plant, monitor, and harvest varieties in their own plots and then share comments on growth patterns, yields, cooking times, and the like with researchers.

This chapter describes an extension of this evaluation process initiated by an ISAR/CIAT team in April 1988. As before, researchers visited farmers' fields, eliciting their evaluations of field trials. But Rwandan farmer bean experts—mainly women—were also invited onto the station to critique varietal trials at an earlier stage of the breeding process; to share expertise directly with station breeders, pathologists, and agronomists; and to select varieties to test in their own fields. Can such "participatory research" techniques lead to the breeding of better varieties—more efficiently—while giving the farmers a greater voice in technology development?

The Setting

Beans are a central item in the Rwandan subsistence regime and are grown by 97 percent of farmers (DeJaegher, 1984). Rwandans purportedly eat more beans (50 kg per capita per year) than any other population in the world (Vis et al., 1972). With average landholdings shrinking to less than 1 ha, population density exploding to nearly 350/km^2 and soil fertility rapidly declining, beans are becoming more, not less, important. Farmers value the crop for its relatively high yields

(c. 800 kg/ha), its rapid maturity, and its high nutritional value. Beans are the "meat" of the countryside, providing 25 percent of the calories and perhaps 45 percent of the protein (ibid.).

Bean research in Rwanda has been relatively vigorous. While isolated varietal selection dates from the 1930s, structured breeding started in 1957 and was intensified in the late 1960s when regional research stations were established (Nyabyenda, 1982). CIAT entered Rwanda in 1983 to further strengthen this national research capacity, and only in the last few years have farmer evaluations begun to be integrated into the national breeding programs. This delay is remarkable for two reasons. First, Rwandan farmers constantly experiment with varieties and have developed stable and productive mixtures. In fact, in several regions, station breeders have a hard time outyielding the local bests (CIAT, 1985). Second, as 95 percent of the population remains rural, it is difficult to find a station researcher whose mother hasn't tended a bean field beside her hut. Yet the gap between station and field research closes ever so slowly.

The Current Structure of Bean Breeding at ISAR, Rwanda

The Overall Schema

The current sequence of bean breeding at ISAR contains several notable features (Figure 6.1). At all stages of on-station testing, beans are selected primarily for yield; disease resistance is an important but secondary consideration, and some high-yielding, disease-susceptible varieties have been released. Further, beans are grown at various station sites for nine to fourteen seasons (five to seven years) before they enter the phase of on-farm testing. Farmer evaluations are incorporated only at this last stage of selection, just before diffusion. The process is time-consuming and station-focused.

This scheme of varietal development is not in theory insensitive either to farmer concerns or to existing agronomic practices. The stated objectives of the on-station breeding program (Nyabyenda, 1986) take account of realistic and variable farming conditions. In addition to promoting high yields and disease and insect resistance, the research agenda aims at the development of varieties that tolerate poorer soils, extreme climatic conditions, and shade; can be grown in association with other crops; and are attractive in color and size to the cultivator and the consumer.

Figure 6.1. Bean selection scheme at ISAR

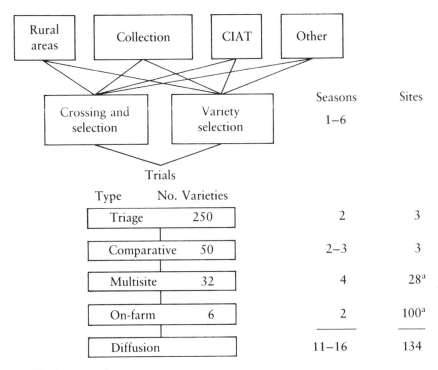

^aNumber varies from season to season.
Source: Modified from Nyabyenda, 1986.

Unfortunately, the reality of the breeding program falls short of its desired ideals for several reasons. Resources of field space and personnel are limited, and as a consequence, priorities lean toward maximizing yields on the more fertile soils. Some of the breeding objectives—for example, tolerance of poorer soils and high yield—prove to be genetically or agronomically incompatible. Further, breeding for performance in crop associations is difficult given the diversity of factors that must be taken into account (crop types, densities, relative positions) and the variety of plant associations farmers practice.

Past Farmer Contributions to National Breeding Programs

The participation of farmers in the Rwandan bean-breeding program has previously meant that scientists take note of farmers' knowledge and assessments only during the period of on-farm testing. Thus, the breeding objectives have been established beforehand and farmers are asked to verify the talents of external experts (see Farrington and Martin, 1988, for fuller discussion of this trend). On-farm trials have been introduced primarily because scientists wish to test their varieties under local farming conditions, and it does not appear that the accompanying farmer evaluations have influenced the development of future varieties.

Scientists at CIAT and ISAR are now asking whether a more full-fledged collaboration between farmers and researchers can alleviate some of the shortcomings of on-station breeding. Farmers have broad and direct experience: they cultivate in a range of soils, in varying associations, and in different seasons. What farmers lack in access to dispersed, replicate trials, they gain from decades of accumulated observation. Further, farmer representatives are in a good position to project which varieties, when released into the countryside, will prove culturally acceptable—and for whom. Farmers' knowledge combined with breeders' talents might produce something better than each expert's isolated efforts.

The Developing Structure: Farmer Participation in On-Station Varietal Selection

Rwandan farmers are now involved in on-station varietal evaluations in the three principal areas of multilocational testing: at high (2250 m), medium (1700 m), and low altitudes (1400 m). At each site, women bean experts judge the fifteen cultivars, sometimes both bush and climbing beans, being tested for forthcoming release. Women make two kinds of evaluations. First they critique varieties as they are growing on-station, and then they predict which varieties would grow best in their own fields, specifying local conditions of soil fertility, crop association, and season. The wisdom of their evaluations is put to the test when participants are permitted to select several varieties for home growing. The salient features of the CIAT-ISAR participatory scheme are sketched below.

Who Evaluates

It is not easy for a researcher to identify farmer bean experts—those who have had considerable success in growing beans, consciously experiment with varieties, and are able to extrapolate beyond the conditions of their own farms. We sometimes ask the community to nominate local experts whom we include in the range of candidates interviewed. We look for thoughtful women who speak clearly and appear not to be intimidated by the presence of men (certainly a concern when the majority of station scientists are male and not conditioned to regard women as "thinkers"). We then visit the women's agricultural plots, looking for evidence of innovation and commitment to good field maintenance ("good" being defined by local agricultural practices).

Finally, if the candidate expresses a strong interest in helping us evaluate varieties, we take the final step—getting the permission of her husband. In Rwanda, *she* may be the "expert," but *he* determines her participation.[1] In our initial set of evaluators, the selected farmers had small to medium holdings (0.3–1.5 ha) with plots of varying fertility. Several had livestock (valued for their manure), and several had nonagricultural sources of income (critical for the purchase of inputs). As representativeness in our sample is essential—that is, choosing farmers with varying wealth and agronomic conditions—we are now also consciously recruiting among the wealthy.

The Evaluative Process

Together and separately, women farmers and station personnel prepare for their encounter. The scientist in charge of the evaluation (who can be a breeder, agronomist, or anthropologist) visits farmers to talk about local evaluative criteria and the conditions under which farmers need to see beans in order to make informed assessments. Because the women have not nurtured the varieties from seed, they have suggested visits at two or three critical points in bean growth: at the flowering and formation of pods, at maturation, and then at harvest.[2] Women feel they can make adequate assessments of varietal performance during a single visit—at the peak of maturation—but they prefer that workers who have maintained the trials be on hand to answer questions on the growth cycle. Station personnel further accommodate the women by marking the trial replicate with named placards and sam-

---

[1] In one amusing incident we learned much about the local assessments regarding varietal expertise. One of the women chosen, an older, stately farmer, fell ill a few days before the on-station evaluation. Her husband suggested he might take her place. "You must be kidding," she said. "I'll send our young daughter first. What do you know about beans?"

[2] As poorer farmers in Rwanda may also eat bean leaves during October and November, when food is short, it was particularly important for some to see the vigor of the foliage.

ples of the seed. Immediately after the evaluation, researchers and farmers discuss the pros and cons of the process and some of the farmers' more significant observations. Follow-up discussions at home help to elicit points farmers may be unwilling to make in public.

The Evaluation Format

Two principles have guided the development of the questionnaire format.[3] First, farmers are given free rein to evaluate the varieties according to their own criteria. Second, the format aims to gather very precise technical information for breeders. The farmers first evaluate each variety separately, commenting on plant architecture, grain, pods, and so on, and assigning a score of 5 (excellent) to 1 (poor). They then select the three best and the three worst varieties, indicating the reasons for their choice. Such an overall assessment highlights the characteristics women deem important (e.g., yield, resistance to rain) but also suggests the relative weighting farmers give to each of the factors. In giving a rounded assessment of their choices, the women may indicate negative aspects of the best varieties as well as the positive features of the worst.

Farmers' On-Station Varietal Assessment

We have thus far completed evaluations in two seasons. I concentrate here on results from the final evaluation during the season ending in January 1989. Several trends emerge:

- When evaluating a variety, farmers consider more than its apparent yield (Table 6.1). Although the bean trials are always laid as sole crops and generally under favorable conditions, farmers judge varieties on such factors as their expected ability to associate with other crops, especially bananas, and their tolerance of stresses likely to be encountered off-station.
- Farmers' assessments do not strictly follow the actual yields of station trials. Varieties selected had among the better yields and those rejected among the lower, but the relationship was not always clear. For example, at the mid-altitude site, the correlations between the mean of the farmers' scores and realized yields was positive but just short of signifi-

[3] The original questionnaire was modified from one already tested during several seasons of on-farm trials (Voss and Graf, 1991). It was altered to permit a more comprehensive evaluation of each variety (a requirement of breeders) and to assure a system of rigorous cross-checks.

Table 6.1. Farmers' evaluations of "best" and "worst" varieties in on-station trials, 1989[a]

Why appreciated	% responses (N = 338)	Why disliked	% responses (N = 219)
High yield	20	Low yield	20
Good architecture		Poor architecture	
Upright	12 ⎱ 21	Sprawling	6 ⎱ 16[b]
Good under bananas	9 ⎰	Poor under bananas	9 ⎰
Performs well under adverse conditions		Performs poorly under adverse conditions	
On poorer soils	5 ⎱	On poorer soils	12 ⎱
In heavy rain	11 ⎰ 24	In heavy rain	17 ⎰ 42[b]
In drought	8	In drought	12
Vigorous pods	10 ⎱ 15	Poor pods	6 ⎱ 11
Vigorous leaves	5 ⎰	Poor leaves	5 ⎰
Early maturing	8	Late maturing	6
Nice color	6	Poor color	2
Large grains	5	Small grains	2

[a] For each of the three best varieties, farmers (N = 35) provided multiple reasons for their choice.
[b] Differences between the sum and its parts result from rounding.

cance (r = 0.51, P > .05) (Table 6.2). As explained above, farmers considered other characteristics as well.

- Farmers are in considerable agreement as to which varieties do well on-station. Accord is often greater, however, on which varieties perform poorly, and why.
- Some farmers select different varieties according to season. Crop associations and especially intensity of rainfall vary over the year. During times of greater stress, the choice of varieties becomes more conservative: farmers choose the sturdier varieties, not necessarily the earliest-maturing or those with greatest yield potential.
- The evaluations clearly highlight regional differences. For example, in an area with little rain, a variety believed to resist drought is rated high, despite its very long cycle. In an area of more abundant rainfall, the same variety is soundly rejected because of its late maturity.
- Farmers are eager to try a wide range of varieties on their fields. However, in all regions, some varieties are rejected categorically. Results suggest that the proportion of varieties rejected is greater in the harsher than in the more permissive environments.

Table 6.2. Farmers' evaluations at mid-altitude multilocational site, *Bush Bean Trial,* 1988–89 (N = 19)

Variety	Yield (kg/4 m²)	Among 3 best (% farmers)	Among 3 worst (% farmers)	Mean[a] score	Comments
PVA 15	3.264	32	*	4.16	Late maturing, poor on less fertile soil
PVA 46	2.950	47	*	4.42	Good yield, fine pods, late maturing, resists disease
K-20	2.436	58	*	4.74	Associates with bananas, late maturing
G11525	2.346	*	57	2.95	Vulnerable to wind, hard to associate
PVA 705	2.323	42	*	4.47	Associates with bananas, good yield, late maturing
Nain de Kyondo	2.106	*	63	3.32	Poor under bananas
RWR 52	2.035	*	*	4.15	
Hatvey 23	1.949	37	*	4.32	
RWR 45	1.927	26	*	4.21	
Kabanima	1.574	*	*	3.84	
PVA 782	1.524	*	*	4.0	
PVA 774	1.517	*	*	3.39	
Kilyumukwe	1.397	*	*	4.0	
ZAA 840/86	1.304	*	*	3.44	
RWR 14	.972	*	79	2.8	Poor yield, poor on less fertile soil, early maturing

Note: * = Farmer accord less than 25%.
[a] Farmers scored varieties from "excellent" = 5, to "poor" = 1.

The Developing Process of Participation

Beyond such specific findings, what has this participatory approach accomplished thus far and what can we anticipate for the future? The assessment of benefits depends on the time frame used and whose perspectives are considered.

The Short Term

Farmers: Women bean experts were unanimously pleased by the station visits. They were exposed to new bean cultivars and were particularly grateful that, this time, they were able to see how varieties performed *before* being expected to plant them in their own plots (hence allowing for better planning on crop associations and planting densities). Further, the bean trials were interesting (especially the climbers, which were new to many), but so were the on-station plots of sorghum and sweet potato. While these farmers have lived within 10–20 km of a research station all their lives, most previously had little idea of what goes on there. Now, some women worry that the varieties they have chosen to test on their home plots might not perform maximally—they feel responsible for and committed to these on-farm trials. In the hopes of being better able to make the link between station and varying farmer conditions, several women's groups in one area are now meeting to select local experts to represent them at on-station evaluations. Station visits are becoming a community issue.

Station scientists: The participation of institute scientists in the on-station evaluations has been encouraging, but commitment has varied in time and intensity. Those most involved have learned about farmer varietal criteria in a comprehensive yet relatively time- and labor-efficient manner. This participatory exercise marks the first time that most station breeders have spoken directly with farmer bean specialists. One breeder was surprised that farmers select varieties according to season (and filled several pages with notes). Another became aware of varietal criteria important to farmers that he had not previously considered—resistance to wind, for example. Certainly each group of experts (on and off the station) still has much to prove to the other. However, useful information has been exchanged—information that had never emerged during several years of on-farm varietal evaluations.

Station administration: In terms of its technical findings alone, the participatory approach has generated sufficient interest at ISAR to be promoted and expanded by the heads of both the Crop Production and the Farming Systems departments. After only two seasons of varietal evaluations they have asked that the method be standardized so as to permit its extension to the twenty-five-odd multilocational sites and that "participatory technique workshops" be arranged to train the needed interviewers.

At the highest level, the administration is enjoying the widespread effects of the publicity. At little cost the research center appears to be taking account of the poor rural population. As these officials have yet to be directly involved in the program and little has been asked of them, their support is still to be tested.

Next Steps

The prospects for further participation within ISAR depend on our ability to find adequate responses to a number of concerns. Widespread skepticism still prevails regarding farmers' ability to select promising varieties for home fields on the basis of what they have observed on-station. This premise is being tested by comparing the on-farm performance of varieties chosen by farmers with those chosen by breeders. To date, we have completed a single full cycle of the participatory exercise: farmers have judged varieties on-station, selected several for home use, and planted and harvested their own trials. Although the sample size is very small—six farmers with eleven varieties—the first results are encouraging. Seventy-three percent of the varieties chosen by farmers performed better than their local mixtures. In the past, varieties selected by ISAR breeders have outperformed the local mixtures in about 45 percent of all on-farm trials.[4]

Operationally, it is far from clear how breeders are to reconcile the two very different data sets with which they now have to cope: quantitative yield data from over twenty-five sites, on the one hand, and, on the other, about seventy farmer evaluations covering aspects of yield, disease, plant architecture, aesthetic characteristics, associability, and more. How much weight should farmers' evaluations be given in selection when the number of farmers involved is relatively small and they judge the variety in only one replicate at a single site?

We have far to go in giving farmers the means to make decisions that can hold their own against the conventional data set. We are experimenting with the advantages of inviting knowledgeable and representative evaluators to take on a more permanent role: we bring the same group of farmers to station trials in successive seasons. Farmers are gaining experience and confidence and are refining their skills of prediction. A danger, however, lies in "professionalizing" these women experts, turning them into adjuncts of the breeding process and isolating them from the source of their experience. Thus, while striving for continuity in farmer expertise, we have also arranged to rotate partially our evaluators.

---

[4] As this paper was being prepared for publication (August 1990), four participatory cycles had been completed. These early findings have been confirmed and even strengthened (see Sperling, 1990).

What has been most striking to date in the participatory program are the steps that farmers and researchers have made to accommodate each other. The women in one community recently took up a collection to help us pay for their transport and refreshment costs—some of the poorest farmers in Africa spontaneously offering to aid a national and international research center. And the station breeders have altered the design of their trials both to give farmers optimum viewing advantage and to respond to farmers' suggestion that on-station varieties be tested on less fertile as well as fertile plots.

The Longer-Term Potential

There is potential for long-term impact in two principal domains: on-farm experimentation and varietal development.

On-farm experimentation: Besides giving farmers a stronger voice in technology (here, varietal) development, participation in on-station evaluations also appears to have important secondary effects, particularly in *upgrading subsequent on-farm testing—with less outside monitoring.*

Farmers participating in on-station evaluations have installed and tended their own home trials with great care. The plots used are those normally reserved for beans (not the leftover stretch often provided for the researcher's experiment), and trials are laid following local principles for varietal testing. Cultivars are tested pure with crop associations, seed densities, and levels of manure and compost regulated according to the farmer's assessment of bean needs as well as other crop demands. Adjacent checks containing the farmer's bean mixtures are planted under the same conditions.[5] These farmer-designed and farmer-managed experiments generally permit a clear interpretation of results—meaningful for both the farmer and the researcher (see also Ashby, 1987). The researcher-designed and jointly managed on-farm trials characteristic of ISAR's more general varietal testing represent a more heterogeneous lot. They vary both in terms of farmers' interest and farmers' and researchers' ability to understand the outcomes.[6] The problems encountered are familiar to seasoned on-farm

[5] In the initial season, a few farmers elected not to lay their local mixture alongside the new variety plot. In their own terms, they explained that they had no need of a direct physical comparison to evaluate the variety. They had been using the plot for years, through different seasons, varying levels of fertility, environmental fluctuations, and the like. They knew the plot's potential; the "check" (so to speak) was in their heads.

[6] It should be noted that the ISAR-CIAT bean team has itself been working to upgrade

researchers: experiments are laid on marginal fields, they are poorly maintained, and treatments are not comparable. We need to examine more closely these qualitative differences in on-farm experimentation under the two regimes. However, if confirmed, these observations suggest that participatory research (starting on-station) can improve the realism and interpretability of on-farm testing. It can also cut supervisory costs by shifting responsibility to the farmer. The interesting corollary development of farmer research groups (cf. Sperling, 1990) suggests that these economies can extend beyond the individual participant to the wider circles in which the farmer shares experimental procedures and results.

Varietal development: Perhaps more directly, farmers' participation in on-station evaluations should help *to increase the efficiency of varietal testing.* In on-station evaluations, farmer agreement has been high as to which varieties can be eliminated from the testing sequence. Early evaluation by farmers will thus allow us to minimize time wasted testing varieties that no one will adopt.

More broadly, the dialogue between breeders and farmers, and their combined efforts, should lead to the *development of more productive and more acceptable varieties.* Breeders can target farmers' needs from the first stages of crossing and selection. Some CIAT scientists have expressed interest in directly involving farmer experts in these initial steps. Others believe it makes little difference whether farmers are physically present when the parents of the cross are chosen— if their interests are well represented. Indeed, farmers may not feel at ease in the offices and laboratories where selection decisions are made.

Somehow breeder and farmer perspectives have to be reconciled. Formerly, if the variety was adopted by *any* large group of farmers within the country, it was considered a success—in breeder's terms. Farmers, however, have a narrower target. They have to make the variety work on a 0.2-ha field. Many farmer selections are aimed at specific conditions (e.g., a variety to be sown in a short-cycle mixture and on less fertile soil). With farmers' guidance, we can determine how common and distinct these niches are. This kind of information can and should be used in setting breeding objectives.

---

the quality of its on-farm varietal trials and to modify them along local experimentation principles. The system, however, has been implemented with differing skill by collaborating development projects.

One forward-looking breeder in our program suggested an even more far-reaching role for farmers:

> Breeding programmes typically generate a great deal of material, which is gradually reduced until one or very few varieties are finally released. These varieties need to be widely adapted if large-scale seed production is to be worthwhile. Beans are rarely a very promising subject for large-scale seed production. An alternative is to develop seed production in cooperatives of small farmers at a local level, and such small-scale seed production would obviate the need for widely adapted varieties. Farmers could then become involved at an earlier stage in the process of selecting among the large genetic diversity in the breeding programme and varieties could be selected which are more specifically adapted to their conditions. (Davis, 1989)

Such a view is among the more progressive, and there is yet no consensus about the future of the ISAR-CIAT participatory program. Considerable strides have been made thus far toward limited goals. Farmers' knowledge has been solicited, but they have yet to share to any degree in decision-making. The constraints on further movement in this direction are considered below.

Costs

We have considered the benefits of on-station evaluation, but what of the costs? Here, I describe only monetary expenditures additional to those incurred in the course of routine on-farm testing. Some of the organizational preconditions as well as philosophical reorientations necessary to launch such a program are discussed at length elsewhere (Biggs, 1989; Chambers et al., 1989; Sperling, 1990).

The direct material demands of this participatory program are minimal: transport for several groups of farmers to and from a multilocational site twice a season, beverages (optional), and the distribution of 200–400 seeds of improved varieties to each participant.

The greatest costs will be encountered in labor. A social scientist will likely be required to initiate the program, to help in identifying local varietal experts and their selection criteria, and to determine their representativeness. He or she will also have to take on the methodological challenges of developing locally adapted evaluation and follow-up procedures. As well, interviewers (including scientists) will need to be trained in and sensitized to participatory research techniques (Ashby, n.d.). Start-up expenses can be much reduced if one of the scientists, social or biological, already has close contact with and knowledge of neighboring farming communities. Beyond the fundamental steps, there is no reason why a breeder or other biological scientist cannot assume prime responsibility for the continuing program.

The actual evaluative process, which should interest a range of scientists as participant interviewers, will require two half-mornings per season of their time. Whether these efforts should be considered supplemental to the normal job requirements of those working to meet farmers' needs is debatable.

In short, once the program is established, direct monetary costs of participatory evaluation are modest. Costs are further offset if one considers the likely spin-off effects: researchers' increased ability to interact effectively with farmers as well as their heightened awareness of farmers' needs and concerns.

The Politics of Participation

The ISAR-CIAT program in participatory evaluation has made significant strides in a relatively short time. Both scientists and farmers have learned from their contacts, and the local bean experts are enjoying the fruits of their station visits—novel bean seeds. On-farm and on-station farmer evaluations represent a small but important foot in what has long been a closed door.

The dominant attitude of the institute, however, remains, for the moment, one of wariness toward the idea of involving farmers in research. Most of the handful of scientists still aspire to issue guidelines unilaterally to some two million farmers. Until recently, breeders bred, and farmers, voting quietly with their hoes, rarely adopted the "improved" seeds. Now, breeders still decide the *whats* and *whens,* and farmers, voting more vocally on evaluation forms, occasionally adopt "improved" varieties.

It is likely that farmers for some time to come will act as no more than nominal participants in the Rwandan scientific arena. Much impedes a closer collaboration. Scientists, having attained high-level posts, are understandably hesitant to share such platforms, and administrative structures reward output, not adoption. Further, working with farmers can be messy, unpredictable, and without immediately tangible results. Most important, however, farmer-responsive research has yet to be fully internalized as a national priority, and Rwandan farmers, dispersed and overwhelmingly silent, make no demands on their research system.

In the light of such constraints the steps forward at ISAR are all the more remarkable. Farmers are being invited to a growing number of researcher-managed sites, scientists and women bean experts are actively working to improve collaboration, and local farmer research groups are starting to guide their own on-farm testing. The limited ISAR-CIAT program is building important bridges, however shaky, between station-based researchers and farmer experts. The imperative to involve farmers further in key decisions affecting their agriculture has yet to be widely recognized. If and when it is, this participatory experience may suggest a way forward.

Bibliography

Ashby, J. *Evaluating Technology with Farmers: A Handbook.* Cali, Colombia: IPRA Projects; CIAT and The Kellogg Foundation, n.d.

——. "The Effects of Different Types of Farmer Participation on the Management of On-Farm Trials." *Agricultural Administration and Extension,* 25:235–53, 1987.

Ashby, J., C. Quiros, and Y. Rivera. "Farmer Participation in On-Farm Varietal Trials." Working Paper of the CIAT/IFDC Project, 1987.

Biggs, S. "Resource-Poor Farmer Participation in Research: A Synthesis of Experiences from Nine National Agricultural Research Systems." OFCOR Comparative Study Paper no. 3, 1989, ISNAR.

Chambers, R., A. Pacey, and L. A. Thrupp, eds. *Farmer First: Farmer Innovation and Agricultural Research.* London: Intermediate Technology Publications, 1989.

CIAT. "Bean Program." In *1984 Annual Report.* Working Document no. 7. Cali, Colombia: Centro Internacional de Agricultura Tropical, 1985.

——. "Farmer Participation." In *Annual Report, Special Projects.* Cali, Colombia: Centro Internacional de Agricultura Tropical, 1988.

Davis, J. "Breeding for Intercrops (with Special Attention to Beans for Intercropping with Maize)." Paper delivered at the Workshop on Research Methods for Cereal/Legume Intercropping in Eastern and Southern Africa, Lilongwe, Malawi, January 22–27, 1989.

DeJaegher, Y. "Resultats de l'enquête nationale agricole 1984 sur les legumineuses." Kigali, Rwanda: MINAGRI, Service des Enquêtes et des Statistiques Agricoles, 1984.

Farrington, J., and A. Martin. "Farmer Participatory Research: A Review of Concepts and Recent Fieldwork." *Agricultural Administration and Extension Network,* 29:247–64, 1988.

Haugerud, A. "Social Science and the Management and Selection of Agricultural Technology in Rwanda." In J. Moock and D. Groenfeldt, eds., *Social Science Perspectives in Managing Agricultural Technology*. New York: The Rockefeller Foundation, 1987.

Nyabyenda, P. *Synthése des résultats de recherche sur haricot au Rwanda durant les 20 dernières anneés*. Rubona: Institut des Sciences Agronomiques du Rwanda, 1982.

Nyabyenda, P. "Programme de recherche objectifs et methodologies ISAR-Rwanda." In *Production et amélioration du haricot dans les Pays des Grands Lacs*. Gitega, Burundi: Institute de Recherche Agronomique et Zootechnique de la Communauté Economique des Pays des Grands Lacs, 1986.

Sperling, L. "Farmers as Partners in Bean Varietal Research." Paper presented at the Ninth SUA/CRSP and Second SADCC/CIAT Regional Bean Research Workshop: Progress in Improvement of Common Beans in Eastern and Southern Africa. Morogoro, Tanzania: Sokoine University of Agriculture, 1990.

Vis, H. L., C. Yourassowsky, and H. Van der Borght. "Une enquête de consommation alimentaire en Republique Rwandaise." Butare, Rwanda: Institut National de Recherche Scientifique, 1972.

Voss, J., and W. Graf. "On-Farm Research in the Great Lakes Region of Africa." In A. van Schoonhoven and O. Voysest, eds., *Common Beans: Research for Crop Improvement*. Wallingford, UK: C.A.B. International in association with Centro Internacional de Agricultura Tropical, 1991.

7

Management of Key Institutional Linkages In On-Farm Client-oriented Research

Deborah Merrill-Sands, Peter Ewell, Stephen Biggs,
R. James Bingen, Jean McAllister, and Susan Poats

Linking Agricultural Research with Resource-Poor Farmers

The Problem

National agricultural research systems are expected to support agricultural development by producing technologies for a wide range of clients: extension services, development agencies, estate crop farmers, large- and middle-scale commercial farmers, and small-scale, resource-poor farm households. It is in meeting the needs of resource-poor farm households—which represent a major share of the world's

This chapter is a revised and expanded version of our article "Issues in Institutionalizing On-Farm Client-oriented Research: A Review of Experiences from Nine National Agricultural Research Systems," *Quarterly Journal of International Agriculture,* 3(4), 1989. Our analysis has benefited greatly from the comments on earlier versions of this paper and related papers from many of our colleagues in the ISNAR study, in particular Marcelino Avila, Stuart Kean, and Sergio Ruano. We also gratefully acknowledge the comments and insights received from numerous reviewers of the earlier comparative study papers on which this paper is based. In particular, we acknowledge Jacqueline Ashby, Michael Collinson, Matt Dagg, Howard Elliott, Pablo Eyzaguirre, David Kaimowitz, Emil Javier, Peter Rood, Willem Stoop, and Robert Tripp. We also gratefully acknowledge the assistance of Jonathan Sands in editing the final draft of the paper. As always, however, we are solely responsible for any omissions or misinterpretations. The views expressed are not necessarily those of ISNAR.

poorest and most vulnerable people—that national agricultural research in developing countries has faced the greatest challenge and had the least success.

The task of responding to the needs of this client group is daunting; technology development for the diverse and complex conditions in which resource-poor farming households strive to produce is difficult. The technological challenge alone, however, does not account for the lack of progress. Institutional constraints and biases are also responsible. Most research systems in developing countries have been organized to support commercial farmers operating in more favorable and homogeneous agroecological areas. Resource-poor farmers have not traditionally been a primary client group, and their needs for technology have not been adequately addressed.

The Role of On-Farm Client-oriented Research

In the last fifteen years, concerns about persistent rural poverty, malnutrition, unsustainable agricultural production, lack of food self-sufficiency, and swelling urban populations have heightened. In response, many governments have undertaken major initiatives to help resource-poor farm households improve their living standards through more stable and productive farming systems. These development objectives have forced shifts in research policy and priorities as well as changes in the organization and management of research and technology transfer systems. Equally important, they have required building strong linkages within these systems and between each of them and resource-poor farmers.

Many research systems have responded by instituting on-farm research programs of varying scope and intensity, designed to bring research closer to farmers. Numerous methodologies have been developed for this type of on-farm client-oriented research, including "cropping systems research," "farming systems research," "on-farm adaptive research," "farmer-back-to-farmer," "farmer-first-farmer-last," and "farmer participatory research" (Ashby et al., 1987; Byerlee et al., 1980; Chambers and Ghildyal, 1985; Collinson, 1987; Farrington and Martin, 1987; Gilbert et al., 1980; Rhoades and Booth, 1982; Zandstra et al., 1981).[1] They all, however, share some essential features:

[1] We avoid using the term *farming systems research* because it tends to be associated with a specific methodology. *On-farm client-oriented research* is used as a generic term to cover a wider range of on-farm research approaches that share the basic characteristics outlined in the paper.

- They involve farmers, as primary clients, in the research process
- They are designed to complement experiment station research
- They emphasize the diagnosis of constraints and setting of research priorities within the context of the whole farm system
- They adapt and evaluate technologies at the farm level

On-farm client-oriented research has proven particularly useful for research systems striving to produce technologies for resource-poor farmers. It offers specific methods to define client groups and identify their research needs, to conduct adaptive research, and to involve farmers in the research process. In developed countries, such activities are carried out expeditiously by agents outside government research institutions: private agriculture input and services companies, extension services, and, above all, farmers. These farmers have access to necessary information, inputs, credit, and services; they can afford to assume the risks inherent in experimentation; and they are organized and have the power to demand the products and information they need from research.

Such conditions do not apply, however, to resource-poor farmers in developing countries. Their access to scientific information is restricted; they have only a limited capacity to tolerate risk; and they are rarely well organized or powerful enough to bring pressure to bear on public-sector research systems. Under these circumstances, on-farm client-oriented research can provide these clients a voice—a means to influence agricultural research and keep it focused on their priorities and relevant to their farming conditions.

Institutional Challenges in Developing and Sustaining On-Farm Client-oriented Research

Substantial progress has been made in developing creative and productive methods for on-farm client-oriented research. However, provisions for fully integrating on-farm research as a stable component within the research system have been inadequate. The institutional challenge has usually been underestimated. Research managers in developing countries face significant problems in organizing and managing this type of research so that it is sustainable within their research systems (Baker and Normal, 1988; CIMMYT and ISNAR, 1984; Collinson, 1982, 1986; Heinemann and Biggs, 1985; Norman, 1983). Too often, on-farm research efforts have been pushed to the side and have not had the intended impact of bringing research closer to the clients—farmers.

The lesson is clear: sound methods are a *necessary* condition for good research, but they are not a *sufficient* condition. Experience has shown that success in on-farm research depends not only on the effectiveness of the methods but also on the institutional and policy context in which they are applied.

The institutional problems do not stem solely, or even primarily, from resource constraints. Rather, many of these problems result from the special and unfamiliar organizational and managerial requirements of on-farm research focused on resource-poor farmer clients. Institutionalizing this type of research effectively means forging a new approach and developing a new set of research activities that complement and build on existing capacities. This is no small task. It involves establishing new communication channels and cooperative efforts among researchers of diverse disciplines, and between researchers and field staff and farmers. It requires systematically training staff or hiring new staff with the right skills. It requires changes in research planning, programming, and review processes in order to formally incorporate farm-level information. It creates increased demands for operational funds and logistical support for researchers working in the field, often far from headquarters or research stations. It often involves working with one or more donor agencies with divergent priorities and distinct operational and funding procedures. And it often entails developing new ways of working with extension at both field and institutional levels.

If on-farm research is to be integrated and sustained as a productive component of national agricultural research systems, these organizational and managerial challenges have to be tackled head-on and surmounted.

INSAR Study on the Organization and Management of On-Farm Client-oriented Research

In response to requests from research managers, the International Service for National Agricultural Research (ISNAR)[2] launched a ma-

[2] The International Service for National Agricultural Research (ISNAR), based in The Hague, Netherlands, is a member of the Consultative Group for International Agricultural Research (CGIAR). Its mandate is to help national agricultural research systems increase

jor study in 1986 to analyze the organization, management, and in-
stitutionalization of on-farm client-oriented research in national agricultural research systems.[3] The objective was to provide a body of
practical experience that research managers can draw upon as they
strive to make on-farm client-oriented research an integral part of
their research systems. The study identified predictable problems in
implementing on-farm research, diagnosed institutional factors leading to such problems, and provided guidelines for managers on
strengthening the effectiveness and efficiency of on-farm research
through improved organization and management.

The approach was to learn from the experiences of research managers in developing countries. The analysis was built around case studies
of national research systems that had formally integrated on-farm client-
oriented research as a major activity and had at least five years' experience. Nine countries were selected for the study: Ecuador, Guatemala, Panama, Senegal, Zambia, Zimbabwe, Bangladesh, Indonesia,
and Nepal. Table 7.1 provides basic descriptive indicators of the case-
study institutions and their on-farm client-oriented research efforts.

ISNAR and the case-study research systems collaborated in the study.
National teams carried out the research and prepared case studies
using an analytic framework developed by ISNAR (Avila et al., 1989;
Budianto et al., in press; Cuellar, 1990; Faye and Bingen, 1989; Jabbar and Abedin, 1989; Kayastha et al., 1989; Kean and Singogo, 1988;
Ruano and Fumagalli, 1988; Soliz et al., 1989). ISNAR then took the
lead in synthesizing the findings, conclusions, and management lessons to be drawn from the cross-country analysis and case experiences (Biggs, 1989; Bingen and Poats, 1990; Ewell, 1988, 1989; Merrill-Sands and McAllister, 1988; Merrill-Sands et al., 1991). This
chapter, which focuses on managing critical linkages in on-farm research, summarizes some of the study's principal findings and conclusions.

―――――――
their efficiency and effectiveness through improved research policy, organization, and management.
   [3] The study was funded by the government of Italy and by the Rockefeller Foundation
through its Social Science Research Fellowship Program.

Table 7.1. Descriptive indicators of the nine OFCOR studies, 1986

| Case studies | National agricultural research system | | Organization of OFCOR | Years in operation | Scale of OFCOR (scientist-years) | |
	Institutional type	Organization of research program			OFCOR as % of NARS human resources	Size of OFCOR effort
Ecuador	Semiautonomous institute (INIAP)	Regional research stations/commodity programs	Production Research Program (PIP):[a] national program with two coordinators and 10 teams based at regional research stations	9	6	14
Guatemala	Semiautonomous institute (ICTA)	Regional research/commodity programs	Technology Testing Department with 14 field teams in 6 regions and national socioeconomics department with limited regional representation[b]	14	34	65
Panama	Semiautonomous institute (IDIAP)	Commodity programs/regional offices	National OFCOR plan identified target regions where OFCOR is implemented through special FSR projects or part-time on-farm research	7	16	24
Senegal	Semiautonomous institute (ISRA)	Multicommodity departments/regional stations	OFCOR, located within Department of Production Systems Research and Technology Transfer (DRSP),[c] consists of 3 regional teams and the Central Systems Analysis Group	4	13	22
Zambia	Ministry (MAWD)	Commodity and factor programs	OFCOR program with national coordinator and 7 provincial teams at regional stations	6	20	38

Zimbabwe	Ministry (MLARR)	Commodity- and disciplinary-based institutes and stations	OFCOR implemented by 8 research institutes/stations with combined on-station/on-farm research programs, and Farming Systems Research Unit (FSRU) based at central station with two regional teams	6	18	26
Bangladesh[d]	BARI, semiautonomous institute of larger NARS with council	Disciplinary departments/ commodity programs	On-Farm Research Division (OFRD), with central management unit at headquarters and 24 teams deployed through BARI's network of regional stations, has official mandate for on-farm research; consolidation of previous OFCOR efforts	9[e]	12	104
Nepal[f]	I. NARS: ministry	I. Commodity programs/ disciplinary departments	I. Farming Systems Research and Development Division (FSR&DD) with 6 FSR sites, supported by Socio-Economics Research and Extension Division (SERED); Commodity programs with multilocational testing and outreach programs	14[g]	n/a	35[h]
	II. LAC and PAC:[i] externally funded autonomous institutes	II. LAC: multidisciplinary research thrusts; PAC: disciplinary departments	II. LAC and PAC regional institutes with OFCOR as a generalized research strategy			

Table 7.1. Continued

Case studies	National agricultural research system		Years in operation	Scale of OFCOR (scientist-years)	
	Institutional type	Organization of research program		OFCOR as % of NARS human resources	Size of OFCOR effort
Indonesia[f]	Ministry, Dept. of Research (AARD) with multiple institutes and coordinating bodies	Commodity-based regional institutes	Two principal modes of implementation: research institutes which conduct OFCOR as part of regular programs, and OFCOR projects organized at AARD level with staff seconded from multiple institutes	11[j]	n/a	57[h]

[a] Programa de Investigación en Producción.

[b] The Spanish names for these departments are Prueba de Tecnología and Socioeconómica.

[c] Département des Recherches sur les Systèmes de Productions et le Transfert de Technologies en Milieu Rural.

[d] The case study is limited to the Bangladesh Agricultural Research Institute (BARI), the largest of the five institutes coordinated by the Bangladesh Agricultural Research Council (BARC).

[e] Refers to NARS. Several OFR programs with complex histories operate within BARI. The oldest, the On-Farm Fertilizer Program, dates back to 1957. This program was reorganized in the late 1970s, about the same time cropping systems research was established at BARI. The OFRD was not formally consolidated until 1984.

[f] The data refer only to the subcase studies unless otherwise indicated; NARS-wide data are not available.

[g] Cropping/farming systems research was initiated nine years ago. On-farm rice research is fourteen years old.

[h] Represents totals for subcase studies only. Not directly comparable with other NARS-wide data.

[i] Lumle Agricultural Centre and Pakhribas Agricultural Centre.

[j] Refers to NARS. In 1973, multiple-cropping research in the Central Research Institute for Food Crops took on a systems orientation and was renamed cropping systems research (CSR). CSR moved onto farmers' fields in 1975.

Managing Key Linkages in On-Farm Client-oriented Research

The case-study experiences show that the success of on-farm client-oriented research depends heavily on the effectiveness of four sets of linkages:

* Linkages between on-farm and experiment station research
* Linkages between on-farm research and farmers
* Linkages between researchers working in different disciplines or commodities
* Linkages between on-farm research and technology transfer agencies

These links are essential. Without them, on-farm client-oriented research cannot perform its designated roles of strengthening the capacity of research to respond to clients' needs and developing technologies that are relevant and feasible for farmers, particularly resource-poor farmers, to adopt.

Building a Strong Partnership between On-Farm and Experiment Station Research

On-farm client-oriented research could complement work carried out on experiment stations. It cannot contribute effectively if it is carried out in isolation (Baker and Norman, 1988; Biggs, 1983; Biggs and Gibbon, 1984; Collinson, 1987; Fresco, 1984; Gilbert et al., 1980; Harwood, 1985). On-farm and experiment station research have discrete but interdependent roles. When linkages are strong, each provides information and services vital to the success of the other.

Adaptive research is the most widely accepted role for on-farm research, which mobilizes knowledge and technologies from experiment stations and applies them to solving farmers' problems or expanding their production opportunities. To be effective in this role, on-farm research depends on applied research of experiment stations to provide a broad range of component technologies which can be adapted to the specific agroecological, socioeconomic, and farm management conditions of specific client groups (Baker and Norman, 1988; Martinez and Arauz, 1984; Moscardi et al., 1983; Norman, 1982; Norman and Collinson, 1985).

Feeding farm-level information back to experiment station research is a second equally vital, although often less accepted, role. On-farm research is used to define client groups, to diagnose key constraints in their farming systems, and to identify researchable problems. This information must then be funneled back to the priority-setting, planning, and programming processes of the research system. Such feedback increases the relevance of research by helping scientists working on experiment stations to better understand farmers' agricultural practices and to focus their research more on farmers' priorities. It also helps them assess the performance of technologies under realistic farming conditions. At the same time, commodity and disciplinary specialists working primarily on experiment stations can give critical support and advice to on-farm research. In addition to providing laboratory services, they can bring their expertise to bear in diagnosing farm-level constraints (such as soil nutrients, water, or pests), designing experiments, and analyzing and interpreting research results.

Case-Study Experiences. The interdependence of on-farm and experiment station research may seem commonsensical. Nevertheless, the case-study experiences reveal that managers in national research systems have had significant problems forging a productive partnership between research conducted on-farm and that carried out on experiment stations. Encouraging scientists to collaborate through exchanging information, specialized knowledge, and support services has been hard to achieve and sustain (Merrill-Sands and McAllister, 1988).

Some aspects of the partnership have been easier to establish than others. In our experience, linkages have been strongest with respect to mobilizing knowledge and technologies from experiment stations and applying them through adaptive research to solving problems identified at the farm level. Some of the case-study programs have achieved important successes in increasing the productivity of small-farm systems with improved and adapted technologies. For example, in Guatemala, new hybrid maize varieties and improved pest control practices developed through on-farm research approximately doubled yields and economic returns in the major farming system of the La Maquina development zone (Ruano and Fumagalli, 1988). In Nepal, various cropping patterns, based on rice and wheat, developed through the Cropping Systems Program increased yields by 20–80 percent and economic returns by 20–60 percent in irrigated lowland areas. Selected cropping patterns became the basis of a large-scale production program which by 1986, only four years after it was begun, covered 100,000 ha (Kayastha et al., 1989).

Such successes have, however, been most evident in the more favorable agroecological environments for which technologies were available "on the shelf" to be adapted to location-specific conditions. In contrast, the lack of technologies appropriate to more marginal and complex environments or noncrop components of the farming system was cited as a major problem in all the cases.

Hardest to institutionalize have been on-farm research's feedback role and the support role of experiment station programs in providing specialized advice to field researchers. Collaboration in these advisory roles was weak in almost half the cases (Merrill-Sands and McAllister, 1988). This finding, especially in regard to relatively mature on-farm research efforts, is disturbing. It is precisely the pooling of expertise—disciplinary and commodity scientists' specialized knowledge combined with on-farm researchers' in-depth understanding of local conditions and farming systems—that is crucial for producing technologies appropriate to the needs of diverse client groups.

Lessons. The case-study experiences indicate that alert research managers should focus their attention on two key areas when building a partnership between on-farm client-oriented research and experiment station research:

- Turning the potential for conflict into constructive debate
- Promoting the exchange of specialized knowledge and information

Turning conflict into constructive debate: Serious and often debilitating conflicts and misunderstandings between researchers working in applied station-based programs and those working on-farm were reported in the majority of cases, especially in the early years of their on-farm research efforts. The tendency toward conflict is hardly surprising, for the very factors that make on-farm and experiment station research complementary and interdependent also create the potential for conflict.

Differences in research objectives—for example, producing component technologies with wide adaptability across a broad range of environments versus adapting technologies to improve the productivity of a whole-farm system in a specific environment—can lead to very different research agendas and priorities. Similarly, differences in methods, experimental design, types of data collected, modes of analysis, and criteria for evaluating results can provoke fundamental disagreements about research priorities and what constitutes good science and credible research. Two common areas of tension were higher coefficients of variation and higher rates of trials loss typical of on-farm research. There were also often strong differences over using additional criteria to evaluate the performance of technologies, such as returns to scarce labor, greater stability in production, shortening of the hunger period, or simply farmers' assessments.

Such differences can and should lead to constructive debate that can spark innovation and improve the quality and relevance of research (Arnold and Feldman, 1986; Kean and Singogo, 1988; Merrill-Sands and McAllister, 1988; Rhoades et al., 1982). Too often, however, such professional debate degenerates into unproductive institutional infighting or simply studied avoidance.

The case studies show that managers can take several steps to minimize conflict and stimulate constructive debate. They can ensure that on-farm client-oriented research is not perceived as a measure to correct past failures of conventional research in generating technology for resource-poor farmers. Managers can also avoid ambiguity, which fosters conflict. They can work with staff to develop a clear and realistic policy on the expected roles, responsibilities, and products of the two sets of research activities. Most important, managers need to make sure that on-farm research is seen as a research activity, and not simply as an extension or demonstration activity.

Promoting exchange of specialized advice and knowledge: Numerous problems obstruct the exchange of specialized advice and knowledge between researchers working on-farm and those working on experiment stations. These typically center on power conflicts, inadequate opportunities for debate and exchange of information, resource constraints, poor motivation, and perceived inequalities in the status of on-farm and experiment station research.

Collaboration and exchange of advice, while leading to better-quality research, at the same time often reduces researchers' scientific independence and control over their work programs. Such threats can spark conflicts of power and scientific judgment. Managers need to make sure that *advisory* roles, such as the feedback and support roles, are not converted into *supervisory* roles. This only breeds resentment and alienates researchers rather than bringing them closer together.

Researchers need to come together for focused and frank discussions if they are to successfully exchange specialized advice and knowledge. Joint visits to the field appear to be most fruitful, especially when senior researchers participate. Periodic joint meetings to review research results and proposals are a basic requirement. Such meetings are most effective when they are kept small, are focused on a specific subject (e.g., commodity, research problem, or agroecological zone), and involve senior scientists who can make decisions on research agendas and planning of collaborative activities. In the cases we studied, managers rarely paid sufficient attention to providing resources for joint field trips and meetings or to making them work effectively.

Managers cannot simply mandate collaboration; they have to provide leadership and incentives. Researchers need to believe that the quality of their own work will improve through collaboration and that management will reward their efforts to support others' research. Unfortunately, researchers too often see the benefits from collaboration accruing to the institution while they carry the costs in terms of scarce time and resources and hardships of traveling to remote areas.

Finally, no one will act upon advice if they do not respect the source. A prerequisite for on-farm research's key feedback role is that researchers on-station view on-farm research as a legitimate and useful scientific activity. The case studies show that on-farm research is only likely to have legitimacy if it has strong scientific leadership and if researchers working on-farm have levels of research experience and degrees comparable with those working on experiment stations. Yet, in five of the nine national research systems studied, on-farm researchers had lower degree levels and less research experience than the average for research staff as a whole. These were precisely the systems where the feedback was least successful (Merrill-Sands and McAllister, 1988).

A viable on-farm research effort requires managers to be willing to commit the requisite resources. If there are not enough experienced scientists to staff an on-farm research effort, then managers have to make sure that junior researchers working in the field are supported systematically by senior scientists. Only experienced researchers can deal with the complexities of field research and effectively communicate and defend the diverse kinds of information and data generated through on-farm research.

Bringing Resource-Poor Farmers into the Research Process

Innovation, relevance, and efficiency in technology development can be greatly increased by systematically incorporating the client's perspective when defining the agenda for research and evaluating its products (Ashby et al., 1987; Biggs and Clay, 1981; Gamser, 1988; Peters and Waterman, 1984; Von Hippel, 1978). Strengthening the involvement of farmers, particularly resource-poor farmers, in research has been a central objective and responsibility of on-farm client-oriented research programs.

Case-Study Experiences. On-farm client-oriented research efforts have been successful in promoting better understanding of clients' priorities and needs within national agricultural research institutions. All the programs studied identified specific client groups, defined and described their farming systems, and diagnosed key constraints and opportunities for research.

On-farm research efforts have faced problems, however, in developing mechanisms for involving farmers actively throughout the research process. In many cases, farmers' involvement has been limited to answering survey questions and carrying out specific roles in conducting and managing trials. In only a few situations have farmers participated directly in research priority-setting and planning exercises, design of experiments or surveys, review and interpretation of results, or systematic evaluation of technologies (Biggs, 1989; Ewell, 1988). In all cases, researchers and technicians found that organizing and managing effective farmer participation in research was much more challenging than was initially expected.

The objectives and procedures for farmer participation varied greatly among the approximately twenty-five discrete on-farm research programs reviewed (Biggs, 1989). In a few programs, researchers were primarily interested in running tightly controlled experiments under diverse agroecological conditions. Here, farmer participation was not an explicit objective and researchers simply "contracted" land or labor from farmers. In the dominant mode of participation, found in more than half the case-study programs, researchers consulted farmers as clients and users of research products. They diagnosed their problems, with varying degrees of farmer input, and tried to find solutions. This process is much like a doctor-patient relationship. Researchers relied primarily on surveys for gathering information about farmers and their farming systems, and on-farm researchers were commonly charged with representing the farmers' point of view in research-planning and review exercises.

In one-third of the programs studied, researchers actively experimented with and established mechanisms for more intensive and continuous participation. In these cases, farmers and researchers collaborated as partners, pooled knowledge, and shared responsibilities for conducting research. Researchers relied on regular research meetings and consultations to have farmers explain both their current farming practices and their demand for new technology, as well as to monitor the progress of on-going research and evaluate its results. This type of collaborative farmer participation has been used most frequently in programs that required routine monitoring of data on farmers' circumstances, such as livestock and pest management research projects, or in programs undertaking research in complex and poorly understood agroecological zones.

Farmer participatory research is increasingly reported in the literature (Ashby et al., 1987; Chambers and Ghildyal, 1985; Farrington and Martin, 1987; Gamser, 1988; Lightfoot et al., 1988; Maurya et al., 1988; Sager and Farrington, 1988). Here, scientists view farmers as colleagues and systematically identify, support, and draw upon farmers' informal research. Only a few examples of such collegiate research were reported in the cases, however, and their innovations had not been incorporated into the institutions' standard procedures for on-farm research (Biggs, 1989).

Lessons. Comparative analysis of the cases (Biggs, 1989) reveals three key management concerns when trying to strengthen the link between research and farmers:

- Sustaining farmer involvement throughout the research process
- Selecting farmer cooperators in accord with research objectives
- Synthesizing farm-level information and incorporating it systematically into the research process

Sustaining farmer involvement: Sustaining farmers' involvement in on-farm research is important in helping research, as a whole, focus on and address the needs of resource-poor farmers (Ashby et al., 1987; Biggs and Clay, 1981; Chambers and Ghildyal, 1985; Gamser, 1988; Goodell et al., 1982; Matlon et al., 1984; Rhoades and Booth, 1982). Relevant research priorities cannot be identified through a single informal survey or rapid rural appraisal. A continuous interchange between researchers and farmers and sufficient time for confidence, trust, and mutual respect to develop is required. Moreover, farmers themselves are innovators, and their experimentation can be strengthened and developed through on-going collaboration with researchers.

Farmers can also play an important role in monitoring research.

Periodic feedback from clients improves the efficiency and effectiveness of research. It can help scientists to identify early on whether the technologies they are developing are appropriate solutions to identified problems. It also helps them to anticipate second-generation problems arising from the use of new technologies. Furthermore, farmers' environments and their resulting farming systems are dynamic. Researchers need to monitor changing circumstances and periodically reassess research priorities in light of farmers' evolving needs and demands.

In spite of these advantages, many of the on-farm programs studied had difficulties incorporating new information from farm-level research after the initial flush of creative energy and enthusiasm of the diagnostic research stage. Attention to farmer participation often waned, and methods and procedures for eliciting information from farmers were often reduced to rather perfunctory techniques.

Regular research meetings with farmers have been used to overcome these problems in Bangladesh, Guatemala, Indonesia, Nepal, Panama, Senegal, and Zambia (Budianto et al., in press; Cuellar, 1990; Faye and Bingen, 1989; Jabbar and Abedin, 1989; Kayastha et al., 1989; Kean and Singogo, 1988; Ruano and Fumagalli, 1988). In these meetings research plans are discussed, on-going research is monitored, and research results are reviewed and evaluated. Perhaps most important, farmers are given more responsibility and power to influence the research agenda (see also Norman et al., 1988).

For such meetings to be effective, however, the case experiences show that they have to be used explicitly as a research tool. Just as with trials and surveys, they have to be carefully planned and managed to meet specific objectives. The purpose has to be clearly defined, participation carefully organized, and conclusions recorded and reported in a timely manner. Such meetings are easier to organize when on-farm trial sites are clustered in villages.

Selecting farmer cooperators: Surprisingly, in light of the attention devoted to farmer selection in farming systems and on-farm research methodologies, most of the programs studied had not given high priority to developing systematic and defensible methods for selecting farmer cooperators (Biggs, 1989; Ewell, 1988).

Researchers often had developed formal criteria for selecting farmers but had not applied them in the field. When faced with problems of contacting farmers, getting trials planted, or reporting results on time, researchers often reverted to selecting farmer cooperators on an ad hoc basis. They simply accepted suggestions of extension agents, took volunteers at meetings, or delegated selection to junior field staff with minimal guidance.

Such ad hoc selection has been shown to bias samples considerably in favor of male, wealthy, politically active farmers (Kean and Singogo, 1988; Soliz et al., 1989; Sutherland, 1986a, 1986b). Unsystematic selection procedures jeopardize the quality and utility of the farm-level information being fed back into the research system. The legitimacy of on-farm research results can be easily challenged, and the ability to extrapolate these results to a larger group of farmers can suffer.

Certainly researchers confront a constant tension in selecting farmers. On the one hand, they need farmers who are representative of an identified client group in order to ensure the relevance of their feedback. On the other hand, they need research-minded farmers who are good collaborators in order to ensure the quality of experimental research results. Procedures for selecting farmers need to accommodate both concerns. Time and resource constraints often make it impossible to use rigorous sampling procedures. Nevertheless, a purposive selection of farmers using explicit criteria based on the objective of the research is a minimal requirement.

Synthesizing farm-level information: The case experiences show that knowledge gained through interaction with farmers has been difficult to synthesize into a form that can be communicated easily to other researchers and retained in institutional memory (Ewell, 1988). In many of the programs studied, the detailed collection of large data sets from trials or surveys, or even the simple comments recorded in field notebooks or taken from meetings with farmers, were not analyzed systematically or incorporated into the research planning and programming process. Moreover, since much of the knowledge researchers and technicians gain while working in the field does not fit into conventional categories for reporting research, it disappears with them if they leave the research system.

Regular meetings with farmers, as described above, are one solution. If conclusions are properly documented, meetings provide a formal means to bring information, knowledge, and feedback from farmers into the research process. A second solution, used in Nepal, Zambia, Zimbabwe, and Bangladesh, is to involve field staff—junior scientists, technicians, and field assistants—in annual research-planning and review meetings. In this way, the knowledge they have gained through frequent informal discussions with farmers can be brought to bear in preparing research agendas (Avila et al., 1989; Jabbar and Abedin, 1989; Kayastha et al., 1989; Kean and Singogo, 1988).

Developing Interdisciplinary Systems Research

Applying an interdisciplinary systems perspective in research is a basic function of on-farm client-oriented research. The clients of on-farm research are resource-poor farm households which operate complex production systems, involving a range of crops and animals, under diverse agroecological and socioeconomic conditions. The development of improved technology for these clients requires a systems perspective and a broader range of research activities than the conventional limits of any single discipline (Byerlee and Tripp, 1988; Collinson, 1987; Harwood, 1979; Norman, 1982; Norman and Collinson, 1985; Rhoades et al., 1982; Zandstra et al., 1981).

Case-Study Experiences. All the programs reviewed put together interdisciplinary teams for at least an initial diagnostic survey or priority-setting exercise. This role for interdisciplinary research is widely accepted. It has proven much more difficult, however, to maintain the broad systems and interdisciplinary perspective in subsequent field research (Byerlee and Tripp, 1988; Ewell, 1988).

Managers in the case studies used very different strategies for organizing ongoing interdisciplinary systems research after the diagnostic stage (Ewell, 1988). Given the strong emphasis on interdisciplinary research in farming systems research methodologies, we were surprised to find that almost half of the on-farm research efforts reviewed deployed individual technical scientists to carry out on-farm research. These researchers were usually agronomists or animal scientists who had received extra training in basic techniques of partial budgeting analysis and on-farm trial methods. They were supported, with varying degrees of commitment, by specialists in both the natural and social sciences.

The remaining programs deployed multidisciplinary on-farm research teams of varying size and composition. Approximately 15 percent had small field teams of an agronomist or animal scientist paired with an agricultural economist. One-third of the programs mounted large farming systems teams including specialists from such additional disciplines as agricultural engineering, agroforestry, nutrition, and anthropology. Such large teams generally must rely on external support in terms of both staffing and operational funds. They can rarely be sustained with national resources alone, at least in small or medium-sized national research systems.

On-farm client-oriented research has clearly brought a social science perspective into field research. But contrary to conventional wisdom, social scientists did not dominate this line of research in the cases studied. They played key roles in launching almost all of the programs studied, but their participation tended to diminish as on-farm research moved from diagnosis into design and testing of technological solutions. For example, in two countries with longer-term on-farm research efforts—Ecuador and Panama—social scientists had completely disappeared from field research at the time of our study. On average, social scientists represented less than 20 percent of the scientific staff in the case-study programs, all of which had at least five years of field experience (Bingen and Poats, 1990; Ewell, 1988).

Lessons. The challenges facing managers striving to develop interdisciplinary systems research are threefold:

- Moving from multidisciplinary to interdisciplinary research
- Maintaining social scientists' inputs
- Sustaining a broad, client-oriented research agenda

Moving from multidisciplinary to interdisciplinary research: The on-farm client-oriented research efforts studied have been successful in developing *multidisciplinary* research. They have had difficulties, however, in developing effective *interdisciplinary* research; that is, research in which scientists from various disciplines are involved in a research program which is mutually planned, executed, and evaluated (Bingen and Poats, 1990; Ewell, 1988). Simply posting researchers from different disciplines to the same team or program does not guarantee collaboration. Scientists tend to retreat into their own disciplinary research agendas. Research management processes are required to bring disciplinary or commodity specialists together to jointly monitor field activities, review research results, and program the following season's work.

The case experiences revealed three other elements critical for building productive interdisciplinary systems research. Strong scientific and team leadership is essential. The on-farm research leader must have solid field experience, training in systems research, and a working knowledge of and respect for the disciplines represented on the team. Second, interdisciplinary research requires experienced scientists who are sufficiently confident in their disciplinary skills that they can afford to be open to contributions from other fields and learn new analytic methods. Last, incentives are needed to encourage scientists to adjust their research agendas in response to the work of their colleagues and to interpret their results in terms of the broader objectives of the team effort.

The often undervalued skill of team building is also critical. Interdisciplinary collaboration is easier to foster in smaller teams with clearly focused mandates. Such teams can be augmented as needed by specialists such as plant pathologists, sociologists, or soil chemists. This approach has worked well in Guatemala and Zambia (Kean and Singogo, 1988; Ruano and Fumagalli, 1988). In many national research systems, small teams are the only option. Research systems with significant resource constraints simply cannot sustain large field teams with numerous disciplinary specialists.

Maintaining social scientists' input: In most of the research institutes reviewed, specific roles for social scientists—primarily agricultural economists but in some cases anthropologists and sociologists—have been institutionalized. These roles concentrate on the early stages of the on-farm research process: research planning, designing and organizing diagnostic surveys and analyzing results, and developing methods for working with farmers. Their roles in performing routine economic analysis of experimental results and in monitoring adoption have also been quite widely accepted.

Evidence from the cases and the literature indicates that if social scientists are to make a broader contribution throughout the research process, they must satisfy several conditions. They must be active in the field, have a good technical understanding of agriculture, and be able to collaborate productively with technical scientists (Biggs, 1989; Byerlee and Tripp, 1988; Ewell, 1988; Horton, 1984; Moscardi et al., 1983; Rhoades, 1984; Rhoades and Booth, 1982; Tripp, 1985). Perhaps most important, they have to use flexible and well-focused research methods that provide timely and relevant information for research planning and interpreting results (Byerlee and Tripp, 1988; Ewell, 1988).

Social scientists help to reinforce the client perspective in research. Among the cases studied, social scientists took lead roles in developing sound and practical methods for selecting both farmer cooperators and innovative mechanisms for encouraging productive farmer participation. Those programs without social scientists, or with minimal social science input, were the most vulnerable to losing their focus on clients' needs and priorities, their dynamism, and their systems perspective.

Staffing constraints impeded institutionalizing social science research. Owing to the lack of national scientists with appropriate training, foreign scientists took the lead in establishing on-farm social science research methods in most of the programs studied. Finding nationals of equivalent status and training to replace these foreign scientists often proved difficult. Moreover, competent social scientists, as professionals with a scarce set of skills, tended to be drawn away from field research into higher-level planning and priority-setting functions within research institutes, or drawn out into more lucrative private-sector jobs.

The lack of established posts for social scientists in many public-sector research institutes is a related constraint. This can cause real problems when managers try to institutionalize on-farm research. The Zambian experience shows that it can take considerable time and effort to create posts for social scientists in civil service structures (Kean and Singogo, 1988). In the interim, employment insecurity and lack of civil service benefits can cause social scientists to seek other employment.

Some managers have attempted to overcome staffing constraints by training agronomists in basic methods of socioeconomic analysis, such as survey design and partial budgeting analysis. Without leadership from a senior social scientist, however, use of these methods has been erratic and difficult to sustain. This is not an effective substitute, even under resource-constrained situations, for the continued involvement of trained social scientists in field research.

Sustaining a client-based research agenda: Interdisciplinary systems research and methodological innovation are hard to sustain beyond the diagnostic stage. Without strong scientific leadership and management there is a strong tendency for on-farm research to narrow from broad systems-oriented and client-oriented approaches to routine on-farm testing of component technologies. The case experiences show that methodological stagnation sets in for several reasons: lack of technical backstopping, staff inexperience, and staff instability.

After the initial diagnosis, senior disciplinary and commodity specialists often stop going into the field to support the work of on-farm researchers. Mechanisms to ensure periodic backstopping of field staff are hard to sustain. They depend on operating funds, clear professional incentives, and the active support of senior research managers.

As on-farm research efforts have expanded, often far too rapidly, an increasing share of the on-farm work has fallen on junior scientists or technicians who simply do not have the training or experience to handle complex systems analysis and apply research methods from multiple disciplines. Nor do they have the authority to defend a client-oriented research agenda or the use of diverse kinds of farm-level data which fall outside the conventions of agricultural research conducted under controlled conditions. The case experiences show that junior researchers under pressure from senior colleagues and without the support of a scientific leader tend to retreat into the safe haven of routine technology testing.

In the cases studied, frequent turnover of on-farm research leaders undermined scientific guidance and contributed to methodological stagnation. Furthermore, training programs in on-farm research methods were often unable to stay ahead of the turnover of field staff. On average, only slightly more than half of the researchers working on-farm had received any specialized training in methods for on-farm research with a systems perspective (Bingen and Poats, 1990).

The case-study experiences show clearly that to sustain a dynamic on-farm research program focused on clients' needs, managers have to bring disciplinary and commodity specialists together to plan, execute, and evaluate research. Equally important, they need to bring researchers working on-farm together to share ideas, critique each others' work, consolidate their experience, and build on each others' methodological innovations.

Forging Effective Linkages with Technology Transfer Agencies

On-farm research cannot be expected to have the geographical coverage necessary for widespread impact. For broad verification, evaluation, and the packaging and delivery of technologies to farmers, especially to resource-poor farmers working in diverse environments, on-farm research has to forge good institutional linkages with technology transfer agencies (Ewell, 1989; Kaimowitz et al., 1989). Such agencies are usually governmental extension services or regional development organizations, but they may also include development agencies, nongovernmental organizations (NGOs), producer organizations, or seed or fertilizer companies (Kaimowitz et al., 1989). Linkages with these agencies in some cases also strengthen the flow of information from farmers to research.

Case-Study Experiences. Virtually all the on-farm research programs reviewed made some efforts to strengthen linkages with technology transfer agencies. Nevertheless, this has been a chronically weak
area in implementation (Ewell, 1989). On-farm research efforts need
to focus more on forging and managing links with technology transfer
if they are to reach a wide range of farmers.

Many programs did not give this objective high priority until late
in the research process when they had technology ready for wide-scale
verification and transfer. Even when managers tried to develop stronger
links early on, it proved difficult. Linking with governmental extension services was particularly problematic; entrenched institutional
barriers and status differences impeded effective collaboration. On-
farm researchers in the case studies also tended to view technology
transfer agents as implementors—people to assist them with the
routine tasks of conducting research on farmers' fields, rather than
partners or even clients. They were only rarely involved in determining the on-farm research agenda and priorities or in evaluating results.

The on-farm research efforts we reviewed vary widely in the degree
of integration they have pursued and developed with technology transfer
organizations. The experiences of Guatemala and Zambia fall at opposite ends of the spectrum. Until recently, Guatemala's Instituto de
Ciencia y Tecnología Agrícola (ICTA) made only sporadic efforts to
link with the government extension service, working on the assumption that technology developed in response to farmers' needs and
adapted through on-farm research would diffuse spontaneously. This
approach has met with success with seed technologies in the more
favorable environments but has been less effective for reaching poorer
farmers in the more complex and marginal hill environments (Ruano
and Fumagalli, 1988).

In contrast, managers in Zambia began from the first to develop
links between on-farm research and extension. Formal mechanisms
were put into place at various levels in the administrative hierarchy,
including provincial coordinating committees and seconding extension field staff and professionals to regional research teams (Kean and
Singogo, 1988).

Lessons. Analysis of the case-study experiences revealed four simple, but not simplistic, lessons for developing stronger linkages between on-farm research and extension agencies (Ewell, 1989). Research managers need to:

- recognize that on-farm research cannot substitute for extension
- move beyond informal field-level cooperation
- develop a partnership with technology transfer organizations
- build linkage mechanisms at multiple levels of the administrative hierarchy

Recognizing that on-farm research cannot substitute for extension: The case-study experiences argue forcefully that on-farm client-oriented research alone is rarely sufficient for the effective transfer of technology. Good technology can sometimes sell itself; some improved crop varieties have diffused widely from on-farm experiments, as the experiences in Guatemala, among others, show (Ruano and Fumagalli, 1988). But this is the exception, not the rule, and transfer efforts can still accelerate the diffusion of such technologies. Most technologies require formal transfer efforts because they depend on specialized inputs, intensive training, or collective action (Kaimowitz et al., 1989). The problem is even more difficult for technologies such as storage or pest management methods consisting of concepts and principles that must be reinterpreted and adapted to each new situation (Ewell, 1989; Horton, 1986; Rhoades, 1987). Transfer efforts usually require resources and skills beyond the capacities of a national research organization. Support from technology transfer organizations must be enlisted and effective coordination mechanisms put into place.

In some cases, however, research institutions took on responsibility for transferring technologies precisely because the governmental extension service could not do the job. Obviously, strengthening linkages cannot compensate for fundamental institutional weaknesses. In such cases, research managers have to explore possibilities for collaborating with alternative technology transfer agencies, such as development projects, nongovernmental organizations, farmer cooperatives, or private-sector seed or input companies. Partnerships with nongovernmental organizations can be particularly effective because these organizations usually have a long-term, focused commitment to development in poor rural areas, work closely with farmer organizations, and are less hampered by bureaucratic constraints.

The clear lesson from the case experiences is that on-farm research should not be expected to replace technology transfer efforts. It can, however, provide a sound basis for forging more effective linkages.

Moving beyond informal field-level cooperation: Most on-farm researchers and research managers in the cases studied viewed extension or development project staff as a resource—a broadly distributed network of people in day-to-day contact with farmers. Consequently, researchers relied heavily on developing informal field-level cooperation. They consulted field agents about local farming conditions and practices and enlisted their help with conducting surveys, selecting farmer cooperators, and organizing meetings and field days. Some researchers even tried delegating more routine on-farm research tasks to extension agents. This is a tempting option: it permits more experimentation and wider geographical coverage, and also increases field agents' understanding of on-farm research activities.

Such informal collaboration improves information exchange and can provide valuable logistic support for on-farm research, but such arrangements are difficult to sustain. When additional tasks are simply tacked onto field agents' normal duties (which generally entail much more than just technology transfer), they rarely have the time, training, mobility, or motivation to carry them out effectively. This is a recipe for frustration: field agents resent collaboration with researchers as an additional and unrewarded burden, and researchers quickly become disillusioned with field staff's apparent lack of interest and poor management of trials or surveys (Avila et al., 1989; Gilbert et al., 1988; Kayastha et al., 1989; Kean and Singogo, 1988; Ruano and Fumagalli, 1988; Soliz et al., 1989).

Formal arrangements, if supported by upper-level managers, can foster effective field-level collaboration and provide reliable support to on-farm research. Field agents need to be given clearly defined duties, the time and the means (e.g., transport and operating funds) to carry them out, and the training required to do the job well. It is also crucial that upper-level managers support these arrangements. This approach has been used successfully in a new effort in Guatemala, where the responsibilities of extension agents teamed with researchers are restricted to technology transfer (Ortiz et al., 1991). Similarly, in Zambia, extension agents have been formally seconded to work as technicians for on-farm research teams (Kean and Singogo, 1988).

Formal arrangements can also improve efficiency in technology transfer. Field agents involved throughout the on-farm research process have a greater commitment to and understanding of technologies being developed and tested. Furthermore, their knowledge of local conditions and farming practices can be drawn on systematically in evaluating research results and farmers' assessments of the technologies.

Developing a partnership with technology transfer organizations: Effective linkages with technology transfer agencies can rarely be based on field-level collaboration alone. Complementary mechanisms designed to involve senior and junior staff from research and extension in the joint planning, monitoring, and review of on-farm activities are needed (Ewell, 1989).

A common approach, used in six of the countries, is to set up coordinating committees involving senior and middle-level research and extension managers. Such committees have had a mixed record of effectiveness. In some cases, as in Zimbabwe and Bangladesh, they provided a mechanism for setting common goals, planning and coordinating activities, pooling resources, and exchanging information efficiently (Avila et al., 1989; Jabbar and Abedin, 1989). But in other cases, where they were formed simply to comply with donors' or policymakers' demands for better coordination, such committees were short-lived or existed in name only (Cuellar, 1990; Faye and Bingen, 1989; Ruano and Fumagalli, 1988).

Assigning senior personnel from extension or development organizations to on-farm research is a second promising strategy for bridging the traditional status gap and building a professional partnership. In Zambia, extension professionals are seconded to regional research teams as research-extension liaison officers (Kean and Singogo, 1988, 1991). Such liaison officers can play a key role in summarizing extension agents' reports on farm conditions and problems for researchers, and in packaging research results into a form that extensionists can use in the field. They can also be pivotal in organizing widespread testing of technologies in collaboration with extension. In Nepal, similar responsibilities are carried out by outreach officers in commodity research programs (Kayastha et al., 1989).

In three of the case-study countries—Ecuador, Indonesia, and Nepal—on-farm research staff participated directly in rural development projects (Budianto et al., in press; Kayastha et al., 1989; Soliz et al., 1989). The clear advantages of these arrangements are that researchers and technology transfer agents collaborate closely under a single funding and management structure and researchers are given specific goals for technology development and adaptation. Nevertheless, all of the on-farm programs that collaborated closely with large development projects experienced a difficult tension. On the one hand, they enjoyed the advantage of having direct links with technology transfer and support systems. On the other, they were vulnerable to short-term project goals and strong pressure to produce quick results. Researchers absorbed in day-to-day demands of the projects are easily cut off from their research institutions. Without strong scientific leadership, the link between their adaptive research and the applied research carried out on stations inevitably weakens. Direct collaboration with development projects is most appropriate where technology is available "on the shelf" and ready for local adaptation.

Building linkage mechanisms at multiple levels of the administrative hierarchy: The case experiences revealed that linkages are most successful and sustainable when mechanisms are active at several administrative levels: among technicians in the field, among researchers and extension specialists at regional or provincial levels, and among senior managers at the highest levels. Such mechanisms reinforce one another. Good cooperation at the field level is impossible to sustain unless regular opportunities to meet and work together are actively supported by management. Again, joint goals agreed upon by high-level coordinating committees cannot be realized unless specific operational procedures are worked out at both regional and local levels.

Conclusions

We have seen that each of the four linkages important for the success of on-farm client-oriented research raises distinct management issues and requires different types of intervention. Managers in the cases studied used a wide range of mechanisms for strengthening these critical linkages (see list below).[4] Some mechanisms, such as joint priority-setting and planning exercises, are more formal and are resource, as well as management, intensive. Others that facilitate the exchange of information or encourage informal cooperation among staff are considerably less demanding.

Linkage Mechanisms[4]

- Planning and review processes
    - —Joint problem diagnosis
    - —Joint priority-setting and planning exercises
    - —Joint programming and review meetings
- Collaborative professional activities

---

[4] Based on the typology of linkage mechanisms developed in D. Kaimowitz and D. Merrill-Sands, "Making the Link between Agricultural Research and Technology Users," prepared as a discussion paper for the workshop of the same title held at the International Service for National Agricultural Research (ISNAR), The Hague, Netherlands, November 19–25, 1989.

—Formal collaboration in trials, surveys, and dissemination activities
—Joint decisions on release of recommendations
—Regular joint field visits (e.g., monitoring tours)
—Informal sharing of tasks and responsibilities
—Informal consultations
• Resource allocation procedures
—Formal guidelines for allocating time to collaborative activities
—Specific allocation of funds for collaborative activities
—Resource-sharing arrangements or exchange of services
—Staff rotations between activities
—Secondment of staff
• Communication devices
—Publications, audiovisual materials, and reports
—Joint training activities or seminars involving both groups
• Assigning coordination responsibilities
—Formal assignment of responsibility for coordination to a specific individual or group or creating full-time liaison positions

To forge effective linkages, managers need to combine various types of mechanisms and apply them at different levels of the administrative hierarchy. Which mechanisms are most appropriate and how they should be applied depends on a range of factors, including the type of linkage being developed, the institutional context, managers' objective in building the link, the nature of the technology or research activity involved, and the kinds of linkage problems faced. For example, managers will employ different mechanisms to forge linkages among groups who resent losing their independence than those they use for groups who recognize the need for collaboration but simply do not take the time to work together and share information.

Clearly, no recipe exists for developing and managing effective linkages. Nevertheless, seven principles can be drawn from the management literature and case-study experiences (Arnold and Feldman, 1986; Ewell, 1989; Kaimowitz et al., 1989; Merrill-Sands and McAllister, 1988; Souder, 1980).[5]

• *Linkages have to be actively managed.* No linkage mechanism, however good, can substitute for managers providing leadership and taking an active role in developing and sustaining links.

• *Linkage mechanisms have to be carefully selected and periodically evaluated.* Linkage mechanisms cost time and money. Sometimes

[5] Based on Merrill-Sands and McAllister, 1988; Kaimowitz et al., 1989; Arnold, 1986; and Souder, 1980.

managers have to make difficult choices in allocating funds and staff time among linkage activities and between these and core research activities. Managers also need to monitor links to guard against mechanisms becoming routinized and losing their utility. Similarly, managers need to be able to respond quickly and flexibly to technical or institutional changes which may create the need for new links or different types of links.

• *Linkages between groups work better when they share a common goal and each group's roles and responsibilities are clearly defined and agreed upon.* Managers have to take the lead in developing a policy which defines the tasks and responsibilities of each group and their relative importance. For such a policy to be effective, the groups have to believe that their assigned roles and responsibilities are legitimate, that their assigned tasks are realistic and feasible, and that the products and services expected of each are necessary for their mutual success. When such a policy is built on staff consensus rather than imposed by directive, it is generally more successful.

• *Linkages cannot be mandated; managers have to create incentives for staff to work together.* Staff members have to feel that the benefits of collaboration outweigh the personal costs. Managers need to make sure that collaborative activities do not become burdensome because they are simply added on top of existing responsibilities. Managers also need to ensure that collaboration and cooperation are explicitly rewarded, that they do not detract from individuals' professional goals, and that staff are compensated for any personal hardships encountered. Something as straightforward as inadequate per diems can quickly erode staff's willingness to participate in joint field trips—a potentially powerful mechanism for developing all four sets of linkages critical for on-farm client-oriented research.

• *Managers have to allocate resources to linkage activities.* If the priority managers give to linkages is ever to be more than rhetorical, they have to demonstrate their commitment concretely. This means explicitly allocating staff time and funds as required to support communication, coordination, and joint activities.

• *Managers have to provide opportunities for staff to interact.* Managers need to promote interaction—both formal and informal. This is common sense, but it is frequently overlooked. This concern is particularly important for on-farm research where scientists and technicians are often working far from headquarters or research stations. Regular interaction among staff facilitates not only the exchange of information but also informal cooperation. Once staff members get to know each other professionally and personally, they are often more motivated to work together. Such informal cooperation can be an important complement to more formal coordination mechanisms when forging linkages among groups.

• *Someone must be responsible for making the links work.* Effective collaboration and coordination among interdependent groups rarely happens spontaneously. The case experiences show that someone has to be responsible and accountable for making the linkages work. Someone has to take the initiative to bring people together for meetings, to organize collaborative professional activities, to ensure that agreements reached between groups are translated into action, to see that information and services are delivered as needed, and to make sure that funds are available as required. If senior managers do not have time to assume such responsibilities themselves, they need to formally assign this task to a member of their staff.

Bibliography

Arnold, H., and D. Feldman. *Organizational Behavior.* New York: McGraw-Hill, 1986.

Ashby, J., C. Quiros, and Y. Rivera. "Farmer Participation in On-Farm Varietal Trials." Agricultural Administration (Research and Extension) Network Discussion Paper 22. London: Overseas Development Institute, 1987.

Avila, M., E. Whingwiri, and B. Mombeshora. *Zimbabwe: A Case Study of Five On-Farm Research Programs in the Department of Research and Specialist Services, Ministry of Agriculture.* OFCOR Case Study no. 5. The Hague: International Service for National Agricultural Research, 1989.

Baker, D., and D. Norman. "A Framework for Assessing Farming Systems Activities in National Settings in West Africa: With Special Reference to Senegal, Nigeria, and Mali. In *Farming Systems in West Africa: Proceedings of the West African Farming Systems Research Network Workshop,* Dakar, Senegal, March 10–14, 1986. Ottawa, Canada: International Development Research Centre, 1988.

Biggs, S. "Monitoring and Control in Agricultural Research Systems: Maize in Northern India." *Research Policy,* 12:37–59, 1983.

———. *Resource-Poor Farmer Participation in Research: A Synthesis of Experiences in Nine National Agricultural Research Systems.* OFCOR Comparative Study Paper no. 3. The Hague: International Service for National Agricultural Research, 1989.

Biggs, S., and E. Clay. "Sources of Innovation in Agricultural Technology." *World Development,* 9(4):321–36, 1981.

Biggs, S., and D. Gibbon. "The Role of On-Farm Research in Strengthening Agricultural Research Systems." In S. Biggs and D. Gibbon, *Agricultural Research to Help Poor People in Developing Countries.* Occasional Paper no. 28. Norwich, U.K.: School of Development Studies, University of East Anglia, 1984.

Bingen, J., and S. Poats. *The Development and Management of Human Resources in On-Farm Client-oriented Research: Lessons from Nine Country Case Studies.* OFCOR Comparative Study no. 5. The Hague: International Service for National Agricultural Research, 1990.

Budianto, J., I. G. Ismail Siridodo, P. Sitpus, D. Tarigans, and A. Mulyadi Suprat. *Indonesia: A Case Study on the Organization and Management of On-Farm Research in the Agency for Agricultural Research and Development, Ministry of Agriculture.* The Hague: International Service for National Agricultural Research, in press.

Byerlee, D., and R. Tripp. "Strengthening Linkages in Agricultural Research through a Farming Systems Perspective: The Role of Social Scientists." *Experimental Agriculture,* 24:137–51, 1988.

Byerlee, D., et al. *Planning Technologies Appropriate to Farmers: Concepts and Procedures.* El Batan, Mexico: CIMMYT, 1980.

Chambers, R., and B. Ghildyal. "Agricultural Research for Resource-Poor Farmers: The Farmer-First-and-Last Model." *Agricultural Administration,* 20:1–30, 1985.

CIMMYT and ISNAR. *Issues in Organization and Management of Research with a Farming Systems Perspective Aimed at Technology Generation,* ed. M. Chang. The Hague: International Service for National Agricultural Research, 1984.

Collinson, M. "Farming Systems Research in Eastern Africa: The Experiences of CIMMYT and Some National Agricultural Research Services: 1976–1981." Michigan State University International Development Paper no. 3. East Lansing: Michigan State University, 1982.

——. "The Development of African Farming Systems: Some Personal Views." *Agricultural Administration and Extension,* 29:7–22, 1986.

——. "Farming Systems Research: Procedures for Technology Development." *Experimental Agriculture,* 23:365–88, 1987.

Cuellar, M. *Panamá: Un estudio del caso de la organización y manejo del programa de investigación en finca de productores en el Instituto de Investigación Agropecuaria de Panamá.* OFCOR Case Study no. 8. The Hague: International Service for National Agricultural Research, 1990.

Ewell, P. *Organization and Management of Field Activities in On-Farm Research: A Review of Experience in Nine Countries.* OFCOR Comparative Study Paper no. 2. The Hague: International Service for National Agricultural Research, 1988.

——. *Linkages between On-Farm Research and Extension in Nine Countries.* OFCOR Comparative Study Paper no. 4. The Hague: International Service for National Agricultural Research, 1989.

Farrington, J., and A. Martin. "Farmer Participatory Research: A Review of Concepts and Practices." Agricultural Administration (Research and Extension) Network Discussion Paper 19. London: Overseas Development Institute, 1987.

Faye, J., and J. Bingen. *Senegal: Organisation et gestion de la recherche sur les systèmes de production,* Institut Senegalais de Recherches Agricoles. OFCOR Case Study no. 6. The Hague: International Service for National Agricultural Research, 1989.

Fresco, L. "Issues in Farming Systems Research." *Netherlands Journal of Agricultural Sciences,* 32:253–61, 1984.

Gamser, M. "Innovation, Technical Assistance, and Development: The Importance of Technology Users." *World Development,* 16(6):711–21, 1988.

Gilbert, E., D. Norman, and F. Winch. "Farming Systems Research: A Critical Appraisal." Rural Development Paper no. 6. East Lansing: Michigan State University, 1980.

Gilbert, E., J. Posner, and J. Sumberg. "Farming Systems Research within a Small Research System: A Search for Appropriate Models." Mimeo, 1988.

Goodell, G., P. Kenmore, J. Latsinger, J. Bandong, C. Dela Cruz, and M. Lumban. "Rice Insect Pest Management Technology and Its Transfer to Small-Scale Farmers in the Philippines." In *The Role of Anthropologists and Other Social Scientists in Interdisciplinary Teams Developing Improved Food Production Technology.* Manila: International Rice Research Institute, 1982.

Harwood, R. *Small Farm Development: Understanding and Improving Farming Systems in the Humid Tropics.* Boulder, Colo.: Westview Press, 1979.

——. "A Winrock Perspective on the Evaluation of FSR/E: Lessons for Indonesian Consideration." Paper prepared for the Workshop on Indonesia's Farming Systems Research, West Sumatra, Indonesia, December 10–13, Sukarami Research for Food Crops, 1985.

Heinemann, E., and S. Biggs. "Farming Systems Research: An Evolutionary Approach to Implementation." *Journal of Agricultural Economics,* 36(1):59–65, 1985.

Horton, D. *Social Scientists in Agricultural Research: Lessons from the Mantaro Valley Project, Peru.* Ottawa: International Research Development Center, 1984.

——. "Assessing the Impact of Agricultural Development Programs." *World Development,* 14:453–68, 1986.

Jabbar, M., and M. D. Zainul Abedin. *Bangladesh: A Case Study of the Evolution and Significance of On-Farm and Farming Systems Research in the Bangladesh Agricultural Research Institute.* OFCOR Case Study no. 4. The Hague: International Service for National Agricultural Research, 1989.

Kaimowitz, D., M. Snyder, and P. Engel. "A Conceptual Framework for Studying Links between Agricultural Research and Technology Transfer in Developing Countries." Linkages Theme Paper no. 1. The Hague: International Service for National Agricultural Research, 1989.

Kayastha, B., S. Mathema, and P. Rood. *Nepal: A Case Study of the Organization and Management of On-Farm Research.* OFCOR Case Study no. 3. The Hague: International Service for National Agricultural Research, 1989.

Kean, S., and L. Singogo. *Zambia: A Case Study of the Organization and Management of the Adaptive Research Planning Team (ARPT). Ministry of Agriculture and Water Development.* OFCOR Case Study no. 1. The Hague: International Service for National Agricultural Research, 1988.

——. "Research-Extension Liaison Officers in Zambia: Bridging the Gap between

Research and Extension." OFCOR Discussion Paper no. 1. The Hague: International Service for National Agricultural Research, 1991.

Lightfoot, G., O. De Guia, Jr., and F. Ocado. "A Participatory Method for Systems-Problem Research Rehabilitating Marginal Uplands in the Philippines." *Experimental Agriculture,* 24(3):301–9, 1988.

Martinez, J., and J. Arauz. "Developing Appropriate Technologies through On-Farm Research: The Lessons from Caisan, Panama." *Agricultural Administration,* 17:93–114, 1984.

Matlon, P., R. Cantrell, D. King, and M. Benoit-Cattin. *Coming Full Circle: Participation in the Development of Technology.* Ottawa: International Development Center, 1984.

Maurya, D., A. Bottrall, and J. Farrington. "Improved Livelihoods, Genetic Diversity, and Farmer Participation: A Strategy for Rice Breeding in Rainfed Areas of India." *Experimental Agriculture,* 24(3):311–20, 1988.

Merrill-Sands, D., S. Biggs, R. J. Bingen, P. Ewell, J. McAllister, and S. Poats. "Institutional Considerations in Strengthening On-Farm Client-oriented Research in National Agricultural Research Systems: Lessons from a Nine Country Study." *Experimental Agriculture,* 27:343–73, 1991.

Merill-Sands, D., and J. McAllister. *Strengthening the Integration of On-Farm Client-oriented Research and Experiment Station Research in National Agricultural Research Systems: Management Lessons from Nine Country Case Studies.* OFCOR Comparative Study Paper no. 1. The Hague: International Service for National Agricultural Research, 1988.

Moscardi, E., V. Cardoso, P. Espinosa, R. Soliz, and E. Zambrano. "Creating an On-Farm Research Program in Ecuador." CIMMYT Economics Working Paper 01/83. El Batan, Mexico: CIMMYT, 1983.

Norman, D. "The Farming Systems Approach to Research." In C. Flora, ed., *Farming Systems in the Field: Proceedings of Kansas State University's 1982 Farming Systems Research Symposium.* Manhattan: Kansas State University, 1982.

——. "Some Problems in the Implementation of Agricultural Research Projects with a Farming Systems Perspective." Farming Systems Support Project Networking Paper no. 3. Gainesville: University of Florida, 1983.

Norman, D., D. Baker, G. Heinrich, and F. Worman. "Technology Development and Farmer Groups: Experience from Botswana." *Experimental Agriculture,* 24(3):321–31, 1988.

Norman, D., and M. Collinson. "Farming Systems Approach to Research in Theory and Practice." In J. Remenyi, ed., *Agricultural Systems Research for Developing Countries.* Canberra: Australian Centre for International Agricultural Research, 1985.

Oritz, R., S. Ruano, F. Jucrez, F. Olivet, and A. Meneses. "A New Model for Technology Transfer in Guatemala: Closing the Gap between Research and Extension." OFCOR Discussion Paper no. 2. The Hague: International Service for National Agricultural Research, 1991.

Peters, T., and R. Waterman. *In Search of Excellence: Lessons from America's Best Run Companies.* New York: Warner Books, 1984.

Rhoades, R. *Breaking New Ground: Agricultural Anthropology.* Lima, Peru: International Potato Center, 1984.

———. "Farmers and Experimentation." Agricultural Administration Unit Discussion Paper 21. London: Overseas Development Institute, 1987.

Rhoades, R., and R. Booth. "Farmer-Back-to-Farmer: A Model for Generating Acceptable Agricultural Technology." CIP Social Science Department Working Paper 1982–1. Lima, Peru: International Potato Center, 1982.

Rhoades, R., R. Booth, R. Shaw, and R. Werge. "Interdisciplinary Development and Transfer of Postharvest Technology at the International Potato Center." In *The Role of Anthropologists and Other Social Scientists in Interdisciplinary Teams Developing Improved Food Production Technology.* Manila: International Rice Research Institute, 1982.

Ruano, S., and A. Fumagalli. *Guatemala: Organización y manejo de la investigación en finca en el Instituto de Ciencia y Tecnología Agrícolas (ICTA).* OFCOR Case Study no. 2. The Hague: International Service for National Agricultural Research, 1988.

Sager, D., and J. Farrington. "Participatory Approaches to Technology Generation: From the Development of Methodology to Wider-Scale Implementation." Agricultural Administration (Research and Extension) Network Paper no. 2. London: Overseas Development Institute, 1988.

Soliz, R., P. Espinosa, and V. Cardoso. *Ecuador: Un estudio de caso de la organización y manejo del programa de investigaciones en finca de productores (PIP) en el Instituto de Investigaciones Agropecuarias.* OFCOR Case Study no. 7. The Hague: International Service for National Agricultural Research, 1989.

Souder, W. E. "Promoting an Effective R&D-Marketing Interface." *Research Management,* 23(4):10–15, 1980.

Sutherland, A. "Extension Workers, Small-Scale Farmers, and Agricultural Research: A Case Study in Kabwe Rural, Central Province. Zambia." Agricultural Administration (Research and Extension) Network Discussion Paper 15. London: Overseas Development Institute, 1986a.

———. "Managing Bias: Farmer Selection during On-Farm Research." *CIMMYT Farming Systems Newsletter,* 26:11–28, 1986b.

Tripp, R. "Anthropology and On-Farm Research." *Human Organization,* 44(2):114–24, 1985.

———. "Farmer Participation in Agricultural Research: New Directions or Old Problems?" Discussion Paper no. 256. Sussex, U.K.: Institute of Development Studies, 1989.

Von Hippel, E. "Users as Innovators." *Technology Review* (January 1978), 1978.

Zandstra, H., E. Price, J. Litsinger, and R. Morriss. *A Methodology for On-Farm Cropping Systems Research.* Los Baños, Philippines: International Rice Research Institute, 1981.

8

Village-Level Studies and Sorghum Technology Development in West Africa: Case Study in Mali

Akinwumi A. Adesina

Introduction

An understanding of farmers' production systems and decision-making behavior is critical for the development of appropriate technology (Chambers et al., 1989; Hildebrand and Poey, 1985). To gain this information, I initiated a village-level study of sorghum-based farming systems in Mali in June 1989 as part of the West African Sorghum Improvement Program (WASIP) of the International Crops Research Institute for the Semi-Arid Tropics (ICRISAT). The study is based on the premise that farmers' production capabilities are determined not just by technical factors but also by the complex economic and sociopolitical factors that make up the larger rural economy (Eicher, 1982; Hildebrand and Poey, 1985). The primary objectives are to understand production systems in sorghum-based cropping systems, to characterize farmer resource endowment patterns and utilization in production, to diagnose resource constraints on increased cereal production, and to test new sorghum technologies on farmers' fields to determine productivity under on-farm conditions. This chapter ex-

The views expressed are the author's and do not necessarily represent those of ICRISAT. Comments from Peter J. Matlon, Timothy O. Williams, and Sara J. Scherr on earlier drafts of the chapter are gratefully acknowledged.

147

amines the farming systems of the upper Niger valley zone and makes suggestions for appropriate technology interventions.

Study Area

The study area is the Operation Haute Vallée (OHV) located in the Sudanian agroecological zone of Mali. Four representative villages were selected and a typology of farms was conducted using farm-type classifications of Mali's farming systems research unit (Kleene et al., 1989). Owing to the importance of animal traction in Mali, the primary classification was based on the ownership patterns of oxen and implements. Two categories of households were defined: hand tillage and oxen tillage. Oxen-tillage households have at least one pair of oxen and either a plow or a multipurpose plow; hand-tillage households use traditional hoes and cutlasses. A random sample of eighty households was selected: forty oxen tillage and forty hand tillage. Data were collected individually and in group discussions with farmers during weekly meetings.

Resource Ownership Patterns

The average farm size for oxen-tillage households varies from 9.8 to 14 ha in the study villages, whereas farm size for hand-tillage households is smaller, ranging from 5 to 8.5 ha. Hand-tillage households are also smaller in membership, with fewer male adults *(actifs)* compared with the oxen-tillage households (Table 8.1).

Ownership of livestock assets and animal traction implements varies from village to village. In general, the oxen-tillage households possess larger numbers of cattle and small ruminants than hand-tillage households. Average equipment ownership among the oxen-tillage households is one plow, one weeder, and one cart. Hand-tillage households depend on hoes *(daba)* and cutlasses; however, they rent tillage implements whenever they are available.

Land-Use Rights and Production Patterns

Landownership in the study villages is based on usufructuary rights. Land cannot be sold. The head of each household is responsible for

Table 8.1. Family structure and resource endowments

Study villages	Kayo HT	Kayo OT	Dibaro HT	Dibaro OT	Ngalamadibi HT	Ngalamadibi OT	Nankilabougou HT	Nankilabougou OT	All villages HT	All villages OT
Family structure										
Men	2.4	5.5	2.5	2.2	3.2	6.9	2.4	2.2	2.6	4.2
Women	2.1	4.3	2.0	4.0	2.8	5.0	2.2	2.4	2.3	3.9
Children	1.0	3.9	1.4	1.4	2.1	6.8	2.7	2.2	1.8	3.6
Total	5.5	13.7	5.9	7.6	8.1	18.7	7.3	6.8	6.7	11.7
Livestock assets										
Cattle	0.4	16.1	0.8	11.4	0.0	6.0	1.0	4.1	0.6	9.4
Goats	0.8	7.5	0.3	7.3	3.0	5.9	1.4	0.7	1.4	5.4
Sheep	0.5	2.6	1.8	8.9	4.0	9.5	0.3	2.7	1.7	5.9
Donkeys	0.3	1.2	0.6	1.8	0.8	1.4	0.0	0.2	0.4	1.2
Oxen	0.1	4.0	0.1	2.4	0.4	2.7	0.0	2.7	0.2	3.0
Total	2.1	31.4	3.6	31.8	8.2	25.5	2.7	10.4	4.2	24.8
Traction equipment										
Plow	0	2	0	1	0	1	0	1	0	1
Weeder	0	1	0	0	0	1	0	0	0	1
Seeder	0	1	0	0	0	0	0	0	0	0
Cart	0	1	0	1	1	1	0	1	0	1
Oxen	0	4	0	2	0	3	0	3	0	3
Total	0	9	0	4	1	6	0	5	0	6

Source: Field data, ICRISAT study villages, 1989.
Note: The values for the equipment are rounded to the nearest number. HT = hand tillage; OT = oxen tillage. The values are averages for farms.

allocating land among household members according to need. Land is neither formally rented nor sold, but farmers in need of extra land can borrow a field from a neighboring household to use during the growing season. The usufructuary rights to land are effective only in the growing season. In the dry season, the itinerant livestock herders are free to search for forage and crop residues on all fields that are not enclosed. This land-use pattern mirrors the observed practice in the southern cotton zone of Mali (Kleene et al., 1989).

There is no landless labor class, as families hold the land in trust for their members, controlling the occurrence of exploitative land relationships or perpetration of land inequalities. While land areas per household tend to favor oxen-tillage households, land-labor ratios do not show large disparities. Land per male adult in animal-traction households is 2.6 ha, compared with 2.2 ha in manual households. Although polarized land inequalities do not exist, differential accessibility to land of varying quality may exist. Some farmers cited lack of good-quality land as a major problem in cereal production.

Two traditional land-use patterns were observed: communal fields and individual fields. Of the 435 fields studied, 80 percent are communal and 20 percent are individual. Decision-making on the communal fields is under the control of the household head *(chef du ménage)*, and production is geared to meet food security needs of household members. Returns from individual fields, allocated by the household head to other sons, wives, or relations living in the *ménage*, are exclusively under the control of the "owner." Individuals within the household are obligated to work on the communal fields during the day but normally return to their individual fields in the late afternoon. The large number of communal fields indicates the importance farmers attach to pooling resources to reduce the risk of crop failure in the precarious environment of rainfed agriculture.

The spatial distribution of land under production (compound land and bush fields) and the nature of the soils determine which crops are planted (Prudencio, 1983; Vierich and Stoop, 1987). Maize and tomatoes are generally cultivated on the more fertile upland fields in the compound, while peanuts, *fonio* ("hungry rice"), *vouandzou*, and sorghum and cowpeas are generally grown on the upland soils in the bush fields. Rice and potatoes are planted in the lowlands where water retention is high and soils are heavier.

Increased cultivation on bush fields and the reduction of the fallow period have implications for sustainable long-term production. However, the usufructuary rights to land present problems to farmers who may otherwise wish to leave their land fallow for a long time or invest in long-term soil conservation technologies. Increasing population densities and the pressure on these marginal lands will necessitate a research priority on new ways to maintain fertility, particularly to control *striga*, a parasitic weed that devastates cereals and is linked to the poor fertility of soils in the bush fields. The risk of losing uncultivated land is high, and farmers may not be willing to invest in long-term soil conservation techniques if their use rights cannot be guaranteed (Noronha, 1988). Therefore, any technology package for regenerating these degraded soils in the bush fields must take traditional land-use patterns into consideration.

Labor Use and Labor Markets

Labor is the most critical input in production, and its availability determines cropping patterns. Labor input is categorized into three groups: male adults, female adults, and children. Intrahousehold labor during the growing season is distributed between the communal fields and the individual fields. The principal fields on which household labor is directed are the communal fields, where the role of women is significant (Table 8.2) and includes hoeing, sowing, weeding, and thinning. However, while women in oxen-tillage households are not involved in the labor-intensive hoeing, women in hand-tillage households spend an average 374 hr/ha in hoeing. The introduction of oxen tillage has therefore substantially reduced the farm labor contribution of women in oxen-tillage households. On average, women in hand-tillage households work 701 hr/ha on communal fields compared with 199 hr/ha for women in oxen-tillage households.

Crops that require the greatest labor inputs are basic cereals—sorghum, millet, and maize. Only after the labor requirements for these basic cereals are met do farmers consider other crops. Labor flow data show that the period with the highest labor demand is the weeding period. For cereals, this extends from the first week in July to the end of September. During the same period, plots of other crops also have to be weeded, leading to labor bottlenecks. Manual households often tend to cultivate more land than they can effectively weed, with the result that several fields may not receive more than one weeding.

Farmers depend on hired labor to supplement family labor. Labor can be hired from other households in the same village or from out-

Table 8.2. Average number of hours worked by households on communal and individual fields by gender and by operation

| | Hand-tillage households | | | | | | Oxen-tillage households | | | | | |
| | Communal fields | | | Individual fields | | | Communal fields | | | Individual fields | | |
	Men	Women	Children	Men	Women	Children	Men	Women	Children	Men	Women	Children
Land clearing	92	—	88	38	19	—	71	—	33	28	30	21
Plowing	151	—	99	26	—	37	213	—	186	70	—	77
Hoeing	178	374	282	—	42	21	—	—	—	—	173	—
Sowing/resowing	177	65	42	36	62	11	151	41	60	90	72	31
Weeding	786	262	252	115	246	18	753	158	294	225	79	113
Fertilization	14	—	—	5	—	—	22	—	3	—	—	—
TOTAL	1398	701	763	220	369	87	1210	199	576	413	354	242

Source: Field data, ICRISAT study villages, Mali, 1989.

side the village. Village labor markets are active because adult male workers who have fulfilled their communal labor requirements (usually two to three days per week) often seek employment on fields of other farmers to meet pressing cash demands. Use of hired labor peaks in the weeding period.

The arrangements for the payment of hired labor vary by village. Generally payment is either in cash, reciprocal work, or grain. Because of imperfect labor markets, influential farmers can get access to labor for work on their own fields at a reduced rate or with only the provision of food. Among the various forms of external labor used is the *association de culture*. These associations can comprise adults, children, women, or men. The remuneration for the use of the services of the association is either in cash or in grains (25–35 kg, payable after harvest).

Two other forms of hired labor, temporary in nature, are daily wage labor from the village and monthly paid labor from outside the village. The monthly external laborers *(navatane)* are housed and fed by a household head. The number of days of required work varies with the season. At the beginning of the season (when there is enough grain in the granaries), the laborer works two days and the remaining days are free for him to look for work on other fields. In the dry season, the days of work by the laborer are increased to three per week. Wages paid to the navatane laborers vary by village. The wage is CFA 5000–10,000 (CFA = franc de la Communauté Finacière d'Afrique) per month at Ngalamadibi village, and CFA 20,000 for the rainy season at Kayo village. However in the latter village the employer does not give any land to the navatane. For daily paid workers, wages depend on the season. During planting and weeding periods the rate is CFA 500–1000 per day. This falls to CFA 150–500 per day toward the end of the season when labor demands are low.

Overcoming Labor Constraints through Animal Traction

Due to the labor problems in the production system, farmers look for labor-substituting technologies, mainly animal traction. Oxen tillage is the major technology used; donkey traction is rare because of their ineffectiveness in the heavy soils in this region. All village households use oxen for land preparation, ridging, weeding, and transportation. The hand-tillage households are at a disadvantage, however, since the oxen are not available for hire during most of the peak periods because the owners are using them in their own fields. Consequently, a majority of the hand-tillage households cannot use oxen for critical operations such as planting and weeding.

Various forms of exchange exist in the rental market for traction implements. The daily rental rate in Nankilabougou village is CFA 2000 per day, while at Ngalamadibi the rate is CFA 3500–4000 per day. In general, the most common method of payment in the animal-traction rental market is reciprocal work. On average, 55 percent of the farmers who borrow equipment use this as the major payment method. Payment by grain was not recorded. In certain cases a manual farmer may request an equipped farmer to corral his oxen team on his field or supply him with manure. Such requests are often rewarded by work on the lenders' fields for varying numbers of days (during the weeding period) depending on the quantity of manure. Manual farmers may also request assistance in field preparation and plowing before the rains. In Ngalamadibi, the daily rental rate for hired oxen traction is five to six days of work by an actif on the field of the owner. Reciprocal work payments for hired traction hours in Kayo average three to four days, and in Nankilabougou they range from two to three days of work during the weeding period. The rates reflect the nonavailability of traction rental during these periods, as traction households are busy on their own fields.

The high dependency of manual households on animal-traction households exacerbates inequalities in the rural production system because it forces the less endowed manual farmers to work in the fields of oxen-tillage households during the critical weeding periods, causing their own weeding operations to suffer. Such dependency relationships create a polarized dual economy, limit production levels of manual households, and accentuate the already skewed income distribution in the rural economy.

Linear programming model analysis performed with the data from the study villages (Adesina, 1992) showed that oxen tillage is highly labor substituting (Table 8.3). The value of total output per adult worker increased from U.S. $147 under hand tillage to $440 with the use of oxen and plow, and to $561 with the use of oxen, plow, and weeder. Cash income (i.e., from crop sales alone) per adult worker increased from $48 under hand tillage to $341 with the use of the oxen and plow, and to $454 with the adoption of oxen, plow, and weeder. Multiyear financial analyses showed that internal rates of returns from use of the full oxen-tillage technology are high (Table 8.4).

Table 8.3. Programming model solutions for alternative adoption patterns of tillage technologies in southwestern Mali

	Average from field with hand tillage	Optimal model solutions		
		Hand tillage	Oxen + plow	Oxen + plow + weeder
Value of total output ($US)	876	678	2022	2581
Value of total output/adult worker ($US)	190	147	440	561
Cultivated area (ha)				
Sorghum	1.8	1.5	2.3	3.0
Maize	0.5	0.4	0.2	0.2
Cowpeas	0.9	0.9	0.0	0.0
Peanuts	0.3	0.2	1.6	2.3
Vouandzou	0.3	0.2	0.0	0.0
Rice	0.9	0.9	0.9	0.9
TOTAL	4.7	4.1	5.0	6.3
Labor substitution with draft power				
Total labor (hours)	2640	2611	2849	3133
Total labor per hectare (hours)	562	643	565	495
Oxen use				
Plowing period			100	100
Weeding period			—	130

Source: After Adesina, 1992.

This may explain farmers' strong demand for animal-traction equipment.

Low Capital Formation and Bias in Credit Markets

Although over 95 percent of the farmers cited a preference for oxen technology, the majority do not have access to credit facilities to finance such purchases. There are few modern credit institutions in this cereal-growing area, largely because of the lack of a major cash crop. Traditional (informal) sources of credit, such as family and friends or private moneylenders, are also very limited because there is very little capital formation in subsistence agricultural systems. The low level of production, lack of modern credit facilities, high transaction costs, and risks faced by moneylenders, coupled with the high premiums faced by farmers, all account for the low use of credit in these cereal-based cropping systems.

Table 8.4 Multiyear financial analysis of the adoption of complete oxen-tillage technology under alternate yield uncertainties, model farm, Sudanian zone, Mali

	Yield risk scenarios		
	Scenario 1 [a]	Scenario 2 [b]	Scenario 3 [c]
Net present value of benefits ($US)	4748	4599	3978
Present value of cost streams ($US)	2311	2311	2311
Benefit-cost ratio	2.1	2.0	1.7
Internal rate of returns	45	44	33

Source: After Adesina, 1992.

Note: The analysis was based on a linear programming farm model developed with data from the villages. To reflect the reality of rainfed cereal production, I generated random yield-adjustment factors using a random number generator in Lotus 1-2-3. These yield-adjustment factors were generated for seven years and used to introduce yield uncertainties. The indicated yield scenarios were analyzed.

The data on variable costs of oxen teams were obtained from the International Livestock Center for Africa (ILCA). These values were based upon village-level data collected from mixed-farming-system households in the semiarid zone of Mali. Costs included in the calculations were payment for vaccines, curatives, and veterinary services; maintenance and herding costs for the animals; and feeding costs for purchasing concentrate supplements, crop residues, and forage. Fixed costs in the calculations include capital costs for purchase of oxen, weeder, and plow. Interest charges on credit for purchase of traction equipments were also incorporated.

[a] Yield-adjustment factors were allowed to range from −10% to 10% of yields in the first year.

[b] Yield-adjustment factors were allowed to range from −20% to 10% of yields in the first year.

[c] Yield-adjustment factors were allowed to range from −20% to −10% of yields in the first year.

The initial cost of establishing full oxen technology is CFA 302,940. Farmers able to purchase oxen technology often divert some of their production capacity into a cash crop such as peanuts to offset the high initial establishment cost of the oxen and equipment. The poor credit position in the zone is a clear example of how the vicious cycle of stifled capital infusion undermines production possibilities and contributes to the low level of output.

Migration: A Consequence of the Low Profitability of Subsistence Farming

Migration of adult male labor is common in all the villages studied. Migration to supplement the poor earnings from rainfed agriculture can be short-term (less than six months) or for more than a year. The two major expenses for which the extra income is needed are taxes

Table 8.5. Reasons for migration

Reasons	Number of farmers				
	Dibaro	Kayo	Nankilabougou	Ngalamadibi	All villages
Low revenue from agriculture	17	2	4	2	25
Earn cash to pay taxes	2	11	12	7	32
Earn cash to buy food grains	2	9	15	5	31
Earn cash for investment in agriculture	—	—	6	—	6
Earn cash for miscellaneous expenses	—	1	4	6	11

Source: Field data, ICRISAT study villages, Mali, 1989.

and additional grain to meet household requirements and social obligations. Only 6 percent of the farmers interviewed cited the need to earn money to invest in agriculture as a factor in migration (Table 8.5).

Very little of the income earned by migrants goes into direct investment in agriculture (9 percent). A large part of migrants' income (30 percent) is used to meet household food security, 12 percent is used for social activities, and 9 percent is used to pay taxes. The finding that migrants' earnings largely aid their households in meeting consumption requirements rather than being used for direct investment in agriculture corroborates findings in other parts of the Sahel (Painter, 1987; Reardon et al., 1988).

The migration of adult male labor for long periods significantly affects production levels in this labor-intensive agricultural system. While most farmers recognize that migrant income is very important for the survival of the household, the majority also feel that migration reduces the supply of labor during critical farming operations. Delayed planting and inadequate weeding often result in lower yields.

The existence of seasonal migration has several implications for development strategies in these subsistence-based farming systems. Among others, it forces some farmers to work their own fields on a part-time basis. Farm income, therefore, is not determined merely by the vagaries of rainfed production systems, but also by the economic climate operating outside the village. Economic instability in the migrant poles affects total farm income and exacerbates the already high-risk production situation in the rural areas.

To reverse this outflux from the farm, it is necessary to increase returns from agriculture. Any proposed technologies to meet this objective should take seasonal labor constraints into consideration. Farmers throughout these study villages plant sorghum together with other crops such as cowpeas and maize, as this gives a higher return to labor than sole-cropping. Improved technologies that increase the yield of profitable cash crops grown along with the cereal staples may provide a more stable income base and help to curb long-term migration.

The next section of this paper uses evidence from the village-level studies to discuss some of the critical factors that should be taken into account in technology development efforts.

Toward Development of Appropriate Technology Interventions

Deemphasize Sole-Crop Sorghum Technologies

Approaches to improve productivity in these systems must be directed at the crops that give high returns to labor. Emphasizing yields per hectare overlooks the fact that labor is the limiting factor for most smallholder farmers. Technologies developed must be directed at increasing returns to labor. Economic analysis of the cropping systems for hand-tillage (Table 8.6) and oxen-tillage households (Table 8.7) shows that of all the crop enterprises, sole-cropped sorghum is one of the least profitable for farmers. Land and labor productivities from other enterprises such as rice, maize and sorghum, and peanut, sorghum, and vouandzou are much higher. Intercropping allows farmers both to increase returns to labor by spreading it over several enterprises and to avoid the risk of crop failure inherent in monocropping. Yet, all sorghum screening by breeding programs in the Sahel has been conducted on sole-cropped sorghums, based on the assumption that farmers would be interested in adopting new varieties which would give them higher yields under a sole-cropped production system.

The results from this village-level study suggest that farmers are not as interested in high-yielding cultivars as they are in early-maturing, yield-stable varieties. Furthermore, sole-cropping sorghum does not give farmers as profitable a return for their labor as intercropping with a legume or cash crop. It is therefore more appropriate to introduce sorghum technologies into the mixed cropping systems with peanuts or cowpeas. Field testing of such sorghum-legume technologies is needed to determine planting dates and cropping densities that will optimize the utilization of water, light, and nutrient efficiency in these intercrops.

Table 8.6. Land and labor productivities for alternative cropping systems, manual-tillage households

| | Village | | | | | | | | Rank based on | |
| | Nankilabougou | | Kayo | | Ngalamadibi | | All villages | | | |
Crops	Ret/ha	Ret/hr	Ret/ha	Ret/hr	Ret/ha	Ret/hr	Ret/ha	Ret/hr	Ret/ha	Ret/hr
Sorghum	36,960	16	30,195	40	—	—	33,577	28	11	14
Sorghum/cowpeas	45,854	61	64,804	71	40,896	66	50,518	66	7	7
Maize	66,000	62	—	—	—	—	66,000	62	5	8
Maize/sorghum	118,875	115	—	—	—	—	118,875	115	3	5
Peanuts/sorghum/vouandzou	185,134	116	—	—	—	—	185,134	116	1	4
Rice	117,018	326	143,550	126	—	—	130,284	226	2	2
Millet/sorghum/cowpeas	—	—	—	—	37,264	49	37,264	49	10	9
Millet/cowpeas	—	—	—	—	43,600	70	43,600	70	9	6
Sweet potatoes	—	—	—	—	117,250	233	117,250	233	4	1
Peanuts/sorghum	—	—	—	—	26,715	45	26,715	45	12	11
Vouandzou/sorghum	—	—	—	—	14,680	47	14,680	47	14	10
Fonio	—	—	—	—	63,591	142	63,591	142	6	3
Millet	—	—	22,935	39	—	—	22,935	39	13	12
Peanuts	—	—	48,360	30	—	—	48,360	30	8	13

Source: Field data, ICRISAT study villages, Mili, 1989.
Note: Ret/ha = return per hectare (CFA); Ret/hr = return per hour (CFA). The wage rate in the agricultural season is CFA 500/day, or CFA 70/hr.
Exchange rate: September 1990, CFA 260 = U.S. $1.00.

Table 8.7. Land and labor productivities for alternative cropping systems, oxen-tillage households

| | Village | | | | | | All villages | |
| | Nankilabougou | | Kayo | | Ngalamadibi | | | |
Crops	Ret/ha	Ret/hr	Ret/ha	Ret/hr	Ret/ha	Ret/hr	Ret/ha	Ret/hr
Sorghum	56,980	77	57,035	75	—	—	57,007	76
Sorghum/cowpeas	53,646	102	74,124	66	70,464	140	66,078	102
Maize/sorghum	137,090	139	—	—	—	—	137,090	139
Peanuts/sorghum/vouandzou	157,552	111	—	—	—	—	157,552	111
Rice	97,614	447	169,488	129	—	—	133,551	288
Millet/cowpeas	—	—	—	—	44,800	89	44,800	89
Sweet potatoes	—	—	—	—	180,250	288	180,250	288
Peanuts/sorghum	—	—	—	—	47,380	115	47,380	115
Vouandzou/sorghum	—	—	—	—	48,832	114	48,832	114
Fonio	—	—	—	—	92,127	204	92,127	204
Maize/millet	—	—	88,035	81	—	—	88,035	81
Peanuts	—	—	109,980	162	—	—	109,980	162
Maize	—	—	53,640	73	—	—	53,640	73
Millet	—	—	41,360	85	—	—	41,360	85
Maize/cowpeas	—	—	94,296	75	—	—	94,296	75

Source: Field data, ICRISAT study villages, Mali, 1989.
Note: Ret/ha = return per hectare (CFA); Ret/hr = return per hour (CFA). The wage rate in the agricultural season is CFA 500/day, or CFA 70/hr.
Exchange rate: September 1990, CFA 260 = U.S. $1.00.

Set Varietal Breeding Objectives Based upon Farmers' Preferences

Farmers cultivate local varieties that are well adapted to the biotic and abiotic stress factors in the zone. The varieties of crops cultivated differ by village zone based upon local tastes and fit within the cropping systems prevalent in each site. Local knowledge of varieties by farmers is an asset that researchers should tap.[1] However, working with local land races is not very popular in breeding programs, despite the fact that these local varieties are well adapted and have desirable food-quality traits. The introduction of new sorghum cultivars, meanwhile, has met with little success (Coulibaly and Coulibaly, 1988). The only sorghum variety that has been widely adopted in Mali is Tiemarifing, a local improved variety. Farmers prefer Tiemarifing because of its early maturity (48 percent), its food quality and adaptability for making different kinds of meals (278 percent), and its high endosperm recovery rates during dehulling (15 percent). Only 7 percent of the farmers cited high yield as the determining factor in the adoption of this variety (Table 8.8). Similar reasons were cited for adopting other food-crop varieties. For millet, early maturity and good taste accounted for 80 percent and 16 percent, respectively, of the reasons for adoption. Because sorghum is grown as a subsistence crop, the demand for high-yielding varieties that do not meet quality and taste requirements will be low.

Farmers' preferences are important in setting breeding objectives for sorghum. Early maturity and yield stability are more important to farmers than high yield in a situation where the persistent risk of drought demands a different planting strategy. The use of long-season, photoperiod-sensitive sorghum varieties will ultimately depend on the rainfall patterns in the Sahel, which, over the last decade, have experienced a continual decline (Dennett et al., 1985). Rainfall in the critical month of August has continued to decline consistently in the Sahel, and the average since 1968 has been 27 percent below the 1931–60 mean. If rainfall is adequate, the long-season cultivars have more advantages in terms of optimization of the use of water in the growing season, but they are also more risky if the rain stops early or a mid-season drought occurs. Evidence across the Sahel shows that more and more farmers are making use of early-maturing cultivars to re-

[1] For a similar argument for rice in the mangrove swamps of Sierra Leone, see Richards, 1986.

Table 8.8. Farmers' preferences for traits of most widely cultivated varieties of sorghum, millet, cowpeas, and peanuts

Crop variety	Variety trait	Percentage of respondents adopting variety because of trait
Sorghum		
Tiemarifing	Early maturity	40
	Adaptability for making different foods and good taste	28
	High endosperm recovery rate	15
	High yield	7
	Resistance to *striga*	2
Keninketélini	Early maturity	90
	Good taste	5
	Drought tolerance and yield stability	5
Kendé	Early maturity	82
	Good taste	9
Tiemantié	Drought tolerance and yield stability	100
Millet		
Tiotioni	Early maturity	55
	Drought tolerance and yield stability	28
Souna	Early maturity	70
	Good taste	13
	Ease of management (i.e., not labor demanding)	17
Peanuts		
47-10	Early maturity	84
	High yield	6
Cowpeas		
Choba	High forage yield	62
	Early maturity	19
	High grain production	10
Maize		
Kabatélini (Magnotélini, Kalossabani)	Early maturity	80
	Good taste	16
	Drought tolerance	4

Source: Field data, ICRISAT study villages, Mali, 1989.

duce the risk of crop failure (Adesina and Sanders, 1991; Deuson and Sanders, 1990; Matlon, 1980; Vierich and Stoop, 1987).

Consider Farmer Risk Attitudes

Farmers' attitudes concerning risk affect the adoption of technologies (Barah et al., 1981; Hardaker and Ghodake, 1984). This is especially true for farmers in drought-threatened environments. Farmers

Table 8.9. Farmer risk attitudes and the adoption of improved sorghum technologies, Sudanian zone, Mali

	Farmer risk-aversion index[a]		
	r = 0	r = 0.2	r = 0.9
Total value of farm output (CFA)	228,712	227,556	225,084
Net cash returns (CFA)	68,722	68,382	66,773
Percentage (%) of sorghum area in			
Improved variety fertilized	13	—	—
Improved variety nonfertilized	—	16	—
Local varieties nonfertilized	87	84	100

Source: Computed from model results.
Note: The improved sorghum varieties analyzed are CSM 388 and Malisor 84-1, underfertilized and nonfertilized options. Yield data on these technologies were based on five years of farm-level tests in the zone of study conducted by SAFGRAD/SRCVO, Institute d'Economie Rurale, Sotuba, Mali.
[a] The methodology is risk programming using the MOTAD approach.

are very cautious in trying new cultivars whose performance depends on the use of chemical fertilizers—a costly input for subsistence farmers (Adesina et al., 1987). Farmers also hesitate to adopt new technologies because they want to avoid income fluctuations that might result from adopting new cultivars.

A programming analysis (MOTAD) of the effects of farmers' risk aversion on technology adoption patterns was conducted using several varieties of sorghum and maize. The results showed that even when cash was not a limiting factor, farmers usually planted 84 percent of their total land in local varieties and only 16 percent in improved varieties. Even then, this class of farmers did not fertilize the improved sorghum varieties because cash outlays can be lost as a result of poor and unreliable rainfall (Table 8.9). The conclusion is that across the various weather regimes, farmers prefer the local land races in terms of yield stability when chemical fertilizers are not applied.

Reexamine Cultivar-led Approaches and Farmers' Perceptions of Production Constraints

After the substantial productivity gains in high-yielding dwarf wheat and rice varieties generated the Green Revolution in Asia, concerns in the research community shifted to giving Africa its own Green Revolution. At the base of this agenda is the "established idea" (Andrews, 1986) that increased food production in developing countries will have to depend on new high-yielding varieties. This presupposes a high demand by farmers for improved varieties, but that may not be the critical base for increased production in semiarid West Africa.[2]

The ranking of production constraints by farmers involved in this village-level study suggest that improved varieties may have a minor role to play. The major factors limiting production, aside from labor, are related to resource management and include lack of animal traction facilities (33 percent); lack of land of good quality (12 percent); lack of fertilizer or effective soil fertility management techniques (18 percent); lack of credit for cereal producers, which limits investment in cash inputs (10 percent); and water stress and drought arising from lack of appropriate soil-water management practices (11 percent). Only 5 percent of respondents indicated that lack of new varieties is the most critical constraint they face.

Poor soil fertility is due to low levels of mineral fertilizer use. The important factors limiting fertilizer use are low cereal-to-fertilizer price ratios, insufficient cash resources, nonavailability of fertilizer on the market when needed, lack of credit for fertilizer purchase, and riskiness of fertilizer use in a rainfed agricultural system.

Organic manure can be used to upgrade soil fertility, but the level of use by farmers is low due to inadequate manure supply and the lack of carts for its transportation to distant fields in the bush lands. The incorporation of crop residues by farmers in the study villages is rare due to (1) the hardness of the soil immediately after the rainy season, which makes incorporation tedious and labor demanding; (2) coincidence of the operation with the period of seasonal migration; (3) lack of knowledge of the practice; (4) competition for crop residues in livestock feeding; and (5) the low body weight and draft power of oxen in the dry season, which severely limits work efficiency. Similarly, mulching is rare because large quantities of mulch are required to make the practice effective in covering the soil. Given the very low levels of use of chemical fertilizers (with no use at all on sorghum and millet), new and cost-effective methods of soil fertility maintenance are urgently needed.

With regard to water management, it will be necessary to develop cost-effective soil and water management practices and other crop management practices that will increase efficiency of water use and

[2] Earlier efforts in West Africa were put into the introduction of high-yielding, fertilizer-responsive sorghum and millet varieties from India. These varieties most often proved to be nonadaptable to the region.

prevent soil erosion, which is increasingly becoming a problem.[3] Therefore, new research efforts in this Sudanian zone may need to shift resources away from breeding and toward other, more critical, problems of farmers. Coulter (1989) pointed out that while about 50 percent of the resources of international agricultural centers go into plant breeding, the existing, and very large, gaps between yields on the experiment station and farmers' fields will be closed not by breeding programs but by improved agronomy. The evidence from this study lends support to this position. This is not to say, however, that breeding programs are unimportant.[4] There is strong evidence that investments in breeding maize hybrids in Zimbabwe and cocoa and oil palms in Nigeria led to on-farm productivity increases in these crops (Coulter, 1989; Eicher, 1982). International research centers have also launched major germ plasm collection efforts to provide genetic diversity in breeding for resistance to multiple stress factors. Despite these efforts, varietal adoption rates for sorghum and millet are very low. The consensus is that breeding work to date has had very limited success in dealing with the low productivity of sorghum and millet in West Africa.[5] As Andrews (1986) asserted, overemphasizing varieties is a restrictive approach.

To be successful, technology development efforts should consider farmers' perceptions of the major limiting factors in production. Resource management issues deserve more emphasis, especially in the Sudanian agroecological zone. This can complement varietal improvement work by reducing the stress factors that severely constrain production. Technology design strategies need to emphasize low-cost methods to improve soil fertility as a precondition to the introduction of varietal technologies. As viewed by farmers, the development of improved land and water management practices and the reduction of soil degradation, water stress, and soil fertility loss are central to the sustainability of production. Even using the currently available cultivars, production could be doubled if resource stresses were removed.

[3] For a good discussion of the available water management options in sorghum- and millet-growing zones of West Africa, see Matlon, 1990 and Nagy et al., 1988.

[4] The international centers' efforts to develop resistances to major pests and diseases such as cassava mosaic virus, headbugs and stem borers in sorghum, maize streak virus, and rice yellow mottle virus are major contributions.

[5] Matlon (1990) argued that with respect to sorghum and millet, a major cultivar-led breakthrough cannot be expected in the near future in the Sahelian and Sahelo-Sudanian zones of West Africa.

Conclusions

This chapter has looked at the sorghum-based farming systems in the ICRISAT study villages in the Sudanian zone of Mali and at the various strategies introduced to increase productivity. The major constraints identified are the availability of labor, water stress, soil fertility, and access to credit. Any technology introduced must consider the subsistence nature of the farming system, the lack of capital formation, and seasonal labor migration.

Research programs that concentrate on breeding high-yielding cultivars without simultaneously improving farmers' agronomic practices are likely to be unsuccessful. Evidence from the study suggests that cost-effective soil and water management and improving other agronomic practices will do more toward realizing the objective of sustainable agriculture than developing exotic cultivars. There is clear complementarity in removing resource stress factors and optimal performance of improved cultivars.

Breeding programs should consider the low-input nature of the agriculture and the risk-averse attitudes of most farmers. Evidence shows that farmers in the Sahel are rapidly moving toward early-maturing cultivars to avoid the risk of drought. Breeding efforts should therefore be directed toward developing a portfolio of cultivars of varying maturities to allow farmers to minimize their risks in adapting to the Sahelian rainfall patterns. Taste and grain quality also play an important role in the adoption of sorghum cultivars.

Because the majority of the farmers would adopt oxen-tillage technology if they obtained credit facilities, efforts should be made to make this valuable labor-substituting input available. To ensure a return on capital investment, a cash crop such as cowpeas or peanuts could be intercropped with sorghum; this has been very successful where it has been introduced.

The study of farmers' production systems is the first step in technology improvement programs. Understanding the linkages between endogenous production factors that affect production is critical for rural development strategies. Farmers' views on major production constraints are very important, and appropriate interventions can certainly result from understanding them.

Bibliography

Adesina, A. A. "Oxen Cultivation in Semi-arid West Africa: Profitability Analysis in Mali." *Agricultural Systems,* 38:131–47, 1992.
Adesina, A. A., P. C. Abbott, and J. H. Sanders. "Ex-Ante Risk Programming Appraisal of New Agricultural Technology: Experiment Station Fertilizer Recommendations in Southern Niger." *Agricultural Systems,* 27(1):23–34, 1987.
Adesina, A. A., and J. H. Sanders. "Peasant Farmer Behavior and Cereal Technologies: Stochastic Programming Analysis in Niger." *Agricultural Economics,* 5(1):21–38, 1991.
Andrews, D. J. "Current Results and Prospects in Sorghum and Millet Breeding." Paper presented at the sixth Agriculture Sector Symposium. Development of Rainfed Agriculture under Semi-arid Conditions, January 6–10. Washington, D.C.: World Bank, 1986.
Barah, B. C., H. P. Binswanger, B. S. Rana, and N. G. P. Rao. "The Use of Risk Aversion in Plant Breeding; Concept and Application." *Euphytica,* 30:451–58, 1981.
Chambers, R., A. Pacey, and L. A. Thrupp, eds. *Farmer First: Farmer Innovation and Agricultural Research.* London: Intermediate Technology Publications, 1989.
Coulibaly, B. S., and O. N. Coulibaly. *Etude sur l'adoption des semences sélectionnées (mil, sorgho et niebe) dans les zones du PFDVS, de l'OMM, de la CMDT, de l'ODIB, de l'OHV et de l'ODIPAC.* Mali: Institut d'Economie Rurale/Division des Etudes Techniques, 1988.
Coulter, J. K., "Research for Agricultural Development in Sub-Saharan Africa." Seventh Ralph Melville Memorial Lecture. Tropical Agriculture Association. *UK Newsletter,* December 1989.
Dennett, M. D., J. Elston, and J. A. Rodgers. "A Re-appraisal of Rainfall Trends in the Sahel." *Journal of Climatology,* 5:353–61, 1985.
Deuson, R. K., and J. H. Sanders. "Cereal Technology Development in the Sahel: Burkina Faso and Niger." *Land Use Policy,* 7(3):195–97, 1990.
Eicher, C. K. "Facing up to Africa's Food Crisis." *Foreign Affairs,* 61(1):154–74, 1982.
Hardaker, J. B., and R. D. Ghodake. "Using Measurements of Risk Attitude in Modelling Farmers' Technology Choices." Economics Group Progress Report no. 60. Patancheru, India: ICRISAT, 1984.
Hildebrand, P. E., and F. Poey. *On-Farm Agronomic Trials in Farming Systems Research and Extension.* Boulder Colo.: Lynne Rienner Publishers, 1985.
Kleene, P., B. Sanogo, and G. Vierstra. *A Partir de Fonsebougou: Presentation, objectifs et methodologie du Volet Fonsebougou (1977–1987).* Amsterdam, Pays-Bas: Institut Royal des Tropiques, 1989.
Matlon, P. J. *Local Varieties, Planting Strategies and Early Season Farming Activities in Two Villages of Central Upper Volta.* Economics Program Progress Report no. 2. Ouagadougou: ICRISAT, 1980.
——. "Improving Productivity in Sorghum and Pearl Millet in Semi-arid West Africa." *Food Research Institute Studies,* 22(1):1–42, 1990.

Nagy, J. G., J. H. Sanders, and H. W. Ohm. "Cereal Technology Interventions for the West African Semi-arid Tropics." *Agricultural Economics*, 2:197–208, 1988.

Norman, D. W., M. D. Newman, and I. Ouedraogo. *Farm Level and Village Level Production Systems in the Semi-arid Tropics of West Africa: An interpretive review of research.* ICRISAT Research Bulletin no. 4, vol. 1. Patancheru, India: ICRISAT, 1981.

Noronha, R. "Land Tenure in Sub-Saharan Africa." Paper presented at the Twentieth International Conference of Agricultural Economists, Buenos Aires, Argentina, August 24–31, 1988.

Painter, T. M. "Migrations, Social Reproduction and Development in Africa: Critical Notes from a Case Study in the West African Sahel." Development Policy and Practice Working Paper no. 7. The Open University, U.K., 1987.

Prudencio, Y. C. "A Village Study of Soil Fertility Management and Food Crop Production in Upper Volta: Technical and Economic Analysis." Ph.D. diss., University of Arizona, Tucson, 1983.

Reardon, T., P. J. Matlon, and C. Delgado. "Coping with Household Food Insecurity in Drought Affected Areas of Burkina Faso." *World Development*, 16(9):1065–74, 1988.

Richards, P. *Coping with Hunger, Hazard and Experiment in an African Rice-Farming System.* London Research Series in Geography 11. London: Allen and Unwin, 1986.

Vierich, H. I., and W. A. Stoop. *On-going Changes in the West Africa Savanna Agriculture in Response to Growing Population and Continuing Low Rainfall.* ICRISAT Cooperative Program for West Africa. Ouagadougou, Burkina Faso, 1987.

9

Labor Patterns in Agricultural Households: A Time-Use Study in Southwestern Kenya

Deborah S. Rubin

Introduction

Introducing new crops or crop technologies into the complex production systems characteristic of rural populations around the world
sometimes has unforeseen and unintended repercussions on household welfare and regional socioeconomic patterns. Increases in household income or total farm productivity may affect household members unequally depending on gender, age, or social position. The
evidence of the diverse effects of development in Africa is documented
in an extensive literature[1] that discusses the complicated relationship
between participation in development projects and, for example, shifts
in household labor patterns (Bryson, 1981; Chaiken, 1990), changes
in established patterns of control over income (Guyer, 1988; Jones,
1986), land use (Dey, 1981; Webb 1989), and, implicitly, time available for child care and altered priorities for allocation of household
resources such as land and labor into both productive and reproduc-

A different discussion of these data appeared in my article "Women's Work and Children's Nutrition in South-western Kenya," *Food and Nutrition Bulletin*, 12(4):268–72,
1990. The assistance of Ellen Payongayong at IFPRI in preparing the tables for this chapter
is gratefully acknowledged. I also thank the International Food Policy Research Institute
and Eileen Kennedy for supporting both my field research in Kenya and my extended stays
in Washington, D.C., to analyze the data.

[1] See, e.g., collections such as Hay and Stichter, 1984; Moock, 1986; Nelson, 1981; and
Davison, 1988.

tive spheres (Kennedy, 1989; Kennedy and Cogill, 1987; Rubin, 1988). Limiting benefits or access to specific farmers may also result in regional economic and sometimes political realignments within a project area (Watts, 1989) or between the project site and nearby areas (Little, 1990).

Quantitative measures of crop yields and income are thus only partial indicators of technological impact. As Berry (1986) suggested, research on both the social organization of production and its technical and price aspects is needed to capture the complex consequences of economic development policies and thus improve the quality of policy-oriented research.

Capturing the diverse microlevel reverberations of changing production patterns was the objective of the time-allocation study presented in this chapter. The research was conducted in a community in Kenya that was part of a sugarcane outgrowers scheme. The focus on time use was intended to identify how different categories of participation in the scheme affected both household-level and intrahousehold-level labor patterns.

The time-use study was part of a collaborative research project in western Kenya involving the International Food Policy Research Institute (IFPRI), the International Centre for Insect Physiology and Ecology (ICIPE), and the Rockefeller Foundation. The project was designed to integrate quantitative and qualitative research methods to determine the effects of commercial sugarcane production on household food security, labor patterns, and health and nutritional status.

An accurate assessment of the impact of agricultural development efforts is particularly important for Kenya given its limited amount of arable land and rapidly multiplying population. The government of Kenya was specifically concerned about the impact of sugarcane schemes because regions with already established cane schemes had experienced a decline in food-crop production and there were indications that children's nutritional status and household food security in those areas had suffered (Williams, 1985).

In addition to a six-hundred-household survey on agricultural practices, food consumption, household expenditures, and maternal and preschooler nutritional status, the IFPRI project[2] included an ethnographic component to identify the social and economic processes through which cash-crop production was influencing household-level

[2]For detailed presentation of survey results, see Cogill, 1987; Kennedy and Cogill, 1987; and Kennedy, 1989.

allocation of resources.[3] The ICIPE research, which included farmer interviews and agronomic trials, aimed at a better understanding of household production practices and priorities to develop more sensitive pest management technologies for increasing household food output and food security.[4] Both research projects sought to identify how households involved in different forms of production varied, if at all, in their household labor patterns as a first step toward tracing the effects of cane farming on other activities such as food-crop cultivation, household chores, and child care.

The time-allocation data provide a detailed picture of, first, the internal household division of labor, especially the differential involvement of men and women in productive work; and, second, the different patterns of labor allocation between households that have entered into contracted sugarcane production and those which have not. When analyzed in an ethnographic context, the data offer a snapshot of the current internal organization of household production and suggest a more dynamic picture of how household production is articulated within larger socioeconomic systems as agricultural commercialization increases.

Interpretation of the time-allocation data reveals that participation in household labor differs according to gender, household involvement in sugarcane production, and social position within the household. Farmers not involved in sugarcane growing put more family labor into food production. In all farming households, men and women have different labor tasks. Men contribute more time to agricultural tasks than women, with male involvement concentrated in land preparation. Men also have more leisure time than women. Women's work includes both agricultural activities and the primary responsibility for household maintenance, leaving little time for leisure or socializing. Women with small children spend less time in farming than other women in the same household.

Examining the tasks performed by each farming group yields other insights about differences in production priorities. Residents of households not growing sugarcane regularly take on wage employment such as weeding cane fields and tending cattle, as well as jobs in small

[3] See Rubin, 1988.
[4] The author was field director of the two projects, The Income and Nutritional Consequences of Cash Cropping in Kenya: The Case of Shifting from Maize to Sugarcane, of the International Food Policy Research Institute (IFPRI); and Food Security and Production Constraints at the Household Level in Western Kenya, of the International Centre for Insect Physiology and Ecology (ICIPE) at the Awendo site in South Nyanza, Kenya, 1985–87.

shops and day labor at the sugar factory. Such temporary wage labor is much less common among both men and women in households that do grow contracted cane, though cane-growing households often run small businesses on the side as well. This finding suggests that movement into sugarcane growing is not equally accessible to all households, and that participation in the sugarcane scheme within the project area is at least partially supported by an unequal system of labor recruitment. Involvement in sugarcane cultivation is therefore not simply the substitution or addition of a single crop; it reflects a different type of productive organization. In addition to the different patterns of time use found in the ethnographic study, the survey revealed that sugarcane farmers have larger farms, more large livestock, and higher incomes (Kennedy and Cogill, 1987; Rubin, 1988), implying that the introduction of cane cultivation has affected the dynamics of both internal household production and community production structures.

Project Background

Fieldwork was carried out in South Nyanza, on the eastern side of Lake Victoria in southwestern Kenya. Commercial sugarcane cultivation was introduced through an outgrowers' scheme in 1978 at the South Nyanza (SONY) Sugar Company, although local varieties of cane had long been grown on a small scale for jaggery production.

Both in Kenya and elsewhere, the introduction of inedible cash crops has sometimes been associated with an increase in preschooler malnutrition,[5] but the precise mechanisms by which changes in farming patterns influence nutritional status remain unclear. The IFPRI research found only small differences in the nutritional status of preschoolers in agricultural households, as measured by caloric intake and anthropometric indicators (Kennedy and Cogill, 1987; Kennedy, 1988). The study nonetheless noted a number of differences in farming patterns between households involved in sugarcane cultivation and those growing other food and cash crops, but there was no obvious connection between these farming patterns and preschooler nutritional status in the short term. These results were obtained during a period of government price protection for cane producers; with re-

[5] Extensive reviews of the contradictory literature concerning the relationship of cash cropping to nutritional status appear in both von Braun and Kennedy, 1986; and Cogill, 1987.

moval of the price supports, the profitability of cane over maize would erode (Kennedy, 1988).

On a longer time scale, however, these shifts in farming patterns represent significant differences in the processes by which households provision themselves and could result in consumption differences. The ethnographic study (Rubin, 1988) detailed a greater diversity in cropping patterns, such as decreased use of intercropping and drought-resistant crops, as well as a different pattern of gender control over food plots between the sugar-farming and non-sugar-farming households.

In many African countries, women's primary role in food production is often the first sphere affected by new crop mixes and crop production technologies. The conceptual framework of the IFPRI study (Kennedy and Cogill, 1987:12; von Braun and Kennedy, 1986:45) proposed that women's time spent in both productive and reproductive activities is linked to their control over income as well as to the nutritional and morbidity patterns of their children. Increasing commercialization of agriculture, by shifting the work load within the household, consequently influences the time women allocate to child care and other activities that have either a direct or an indirect bearing on health and nutritional status. However, the first phase of the investigation into the shifting labor patterns associated with sugarcane cultivation (Cogill, 1987) found that other criteria, such as income attributed to sugar, farm size, and latrine use, were more strongly related to better child nutritional status than was women's use of time.

The time-allocation data used in Cogill's analysis were limited in scope: Only mothers of preschoolers were interviewed, and many types of activities were merged into general categories. As a result, the analysis revealed few differences in the daily activities of these women according to the household's involvement in cane farming. More detailed scrutiny of time use by all household members, including men and other women, might clarify the relationship between time use and child care by indicating whether differences in time use by mothers of preschoolers are greater than first thought, what the time-use patterns by other household members are, and if time-use patterns between farming groups parallel the differences in their cropping patterns.

Methods

Before time use was investigated, information on household com-position was collected for seventy-five households—identifying each household member and his or her age and relationship to the others. Careful identification of each person's position within the household structure proved important in disaggregating the time-allocation material. A variety of household structures was discovered, and sugar-cane-farming households showed a higher rate of polygynous marriages, both historically and during the study, than those of non–sugar farmers. This suggests that cane farmers were wealthier than their neighbors even before participating in the scheme.

In the original survey work, data were collected only on mothers of preschool children, who were asked to state how many hours they had spent in different types of household, agricultural, and other activities during the previous day. Total times were calculated by enumerators during the interview and recorded against a set of precoded activity categories. For this study, a new questionnaire schedule was devised, and slightly different interview techniques were used to obtain greater detail about daily tasks. First, a list of common household tasks was compiled based on observations of people's activities in the household and in town. The questionnaire form separated the day into the segments commonly used and recognized by Luo speakers. Because of the scarcity of functioning clocks and watches in rural households, respondants were not asked to report their activities as aggregated hours but rather according to the sequence of tasks over the course of the previous twenty-four hours. Enumerators were instructed to probe respondents' answers by using local time cues such as the factory shift whistles, school schedules, and bus schedules. The enumerators recorded the tasks described to them without coding the information; coding was done later in the office by one employee, and each sheet was checked by the field director.

These changes afforded several advantages over the more generalized questionnaire schedule: (1) the twenty-four-hour period and Luo time cues helped to structure the responses and provided a check on total time spent; (2) a more complex set of activities engaged in over the day was revealed and resulted in new codes being added; (3) listing tasks in sequence outlined the daily schedule of work; (4) consistent categorization of activities was obtained by limiting the coding to a single person; and (5) interviewing both men and women revealed the division of labor common within households and between co-wives. Important differences appeared between men's and women's activities, in the use of time by different groups of women, and in the use of time between households engaged in cane farming and other agricultural households. Over the course of twelve months, more than eight hundred interviews were conducted about respondants' activities during the previous twenty-four-hour period.

Problems nonetheless remain in time-allocation data collection. Gross (1984) reported that recall studies are generally less accurate than those based on direct observation. There is also a respondent bias toward household members who are found in and around the compound. Accurate weighting of simultaneous activities is difficult. In the questionnaire, if a response indicated activities—for example, child care and food preparation—the total amount of time was divided between each activity.

Results of the Time-Allocation Study

Gender appears to be a primary determinant of time allocation in Luo households of all types, and men and women concentrate their work in different spheres (Table 9.1). The extent to which gender-based tasks differ, however, varies according to the primary economic orientation of the household as a whole. Table 9.2 reveals that time-use patterns in merchant and factory households are significantly different from those in agricultural households.

Time allocation in agricultural households does not differ in statistically significant ways, but it does illustrate qualitatively different patterns. The variations evident in disaggregating the categories underscore the importance of considering intrahousehold as well as household characteristics in measuring project impacts.

Gender remains the major differentiating feature in time use in agricultural households. The frequency of activities shown in Table 9.3 illustrates the general pattern of sex-segregated tasks found in Luo agricultural households. Women's work centers on the daily reproduction of household necessities as well as food-crop farming. Men show little activity in domestic and food-related tasks; they report higher participation in general agricultural and animal care activities and infrequent tasks such as compound repair and carpentry.

Comparison of the disaggregated data in Tables 9.4 and 9.5 shows that while both men and women spend roughly similar amounts of

Table 9.1. Time allocation of women and men in all households (hours/day)

Activity	Women N=432	Men N=180
Domestic	8.53	4.79
Agriculture		
Sugarcane farming	0.11	0.39
All other farming	1.90	2.29
Other income-generating work	0.86	2.08
Other activities		
Sleeping	9.08	8.57
Relaxing	1.91	2.54
Community, travel, socializing	1.61	3.09
TOTAL	24.00	23.75[a]

All households: Sugar farmers, other farmers, merchants, SONY employees, and landless households. *Domestic work:* Food processing and cooking, fetching water and firewood, housecleaning, compound repair, child care, shopping, bathing, and animal care. *Agricultural work:* Land preparation, planting, weeding, fertilizing, and harvesting. *Other income-generating work:* Crafts and other small-scale manufacturing, and wage employment. *Other activities:* The category of relaxing includes sick time; community, travel, and socializing includes churchgoing, visits with friends and relatives, attendance at funerals, hospital or clinic visits, and other travel.
[a] Total not 24 hours because of incomplete responses for some men.

time in personal activities, there is a large difference in their use of leisure and social time. Men sleep fewer hours on average than women but spend over five hours during the day relaxing, visiting with friends, and in other community affairs, compared with less than three hours spent by women in the same activities.

Observation and time-allocation analysis shows both men and women taking part in agricultural work and animal care. Men engaged in food-crop farming 50 percent of the days in the sample; women farmed on 45 percent of the days interviewed. The type of farming work they participate in is different. Men plow and do other tasks of land preparation to a greater extent than women. Both men and women plant, weed, and harvest staple grain crops, but in these tasks women do a greater share of the work. Marketing and processing are primarily the province of women. Women cultivate vegetable fields; men are responsible for tree crops. The pattern of high male participation in farming seen in the time allocation is supported by information collected on males' assistance in female-controlled food plots in another component of the study (Rubin, 1988). Qualitative review of these labor data shows a distinct daily schedule of tasks in addition to the definite division of tasks associated with different gender roles.

Attention only to gender, however, glosses over the differences in

Table 9.2. Time allocation according to primary household economic activity
(hours/day)

Activity	Sugar farmers ♀ = 230 ♂ = 84	Other farmers ♀ = 113 ♂ = 62	Merchants ♀ = 13 ♂ = 13	SONY workers ♀ = 24 ♂ = 3	Landless ♀ = 52 ♂ = 18
Domestic					
Women	8.55	7.89	6.51	12.15[a]	8.10
Men	3.35	3.32	3.69	3.50	2.77
Animal care					
Women	0.11	0.05	0.00	0.00	0.00
Men	1.83	1.79	0.00	0.00	0.04
Cane farming					
Women	0.16	0.08	0.08	0.00	0.03
Men	0.50	0.25	0.67	0.00	0.21
Other farming					
Women	1.89[b]	2.67[c]	.96	0.22[b]	1.21[c]
Men	2.12	3.27[d]	.06[d]	1.17	1.14
Employment					
Women	0.05[e]	0.22[e]	5.37[f]	0.00	1.04[e]
Men	0.95[g]	1.32[g]	7.39[g]	3.17	3.44[g]
Manufacturing					
Women	0.60	0.44	0.00	0.76	0.20
Men	0.05	0.25	0.00	0.00	0.00
Other activities					
Women	12.64	12.65	11.08	10.87	13.42
Men	14.84	13.59	12.19	16.16	16.21
Totals[h]					
Women	24.00	24.00	24.00	24.00	24.00
Men	23.64	23.79	24.00	24.00	23.81

[a] Time spent by women on domestic activities significantly different from that spent in all other household groups.

[b] Time spent in other farming found to be significantly different between sugar-farming women and women in households employed by the sugar factory.

[c] Time spent on other farming found to be significantly different between women in non-sugar households and those in landless households.

[d] Time spent by non-sugar-farming men significantly different from men in merchant households.

[e] Time spent in wage employment by women in landless households.

[f] Time spent in wage employment by women in merchant households significantly different from all other groups.

[g] Time spent by merchant men in employment significantly different from men in agricultural and landless households.

[h] Totals are not all 24 hours because of rounding (men and women) and incomplete responses regarding leisure activities for some men.

Table 9.3. Frequency of activities cataloged for agricultural households from time-allocation interviews by sex

Activity	Women (358 days)	Activity	Men (146 days)
Sleeping	358	Sleeping	146
Eating	350	Eating	143
Food preparation	328	Resting/relaxing	118
Bathing	293	Bathing	118
Housecleaning	259	Animal care	77
Resting/relaxing	226	Food-crop farming	73
Fetching water	197	Travel/transport	58
Food-crop farming	162	Visiting friends	37
Child care	155	Compound repair	18
Travel/transport	91	Food preparation	14
Shopping	59	Sugarcane labor	10
Threshing and grinding	58	Agricultural labor	10
Collecting firewood	55	Church or community	9
Crafts/carpentry	51	Other farming	6
Gathering greens	46	Crafts/carpentry	6
Community or church	43	Visiting hospital	5
Marketing	39	Attending funerals	4
Visiting with friends	27	Marketing	4
Animal care	22	Shopping	4
Visiting hospital	16	Child care	4
Sugarcane farming	13	Wage employment	4
Compound repair	10	Fetching water	3
Attending funerals	7	Housecleaning	3
Agricultural wage labor	5	Collecting firewood	1
Wage employment	4	Grinding/threshing	0
Other farming	2	Gathering greens	0

Note: Eating is not indicated each day because meals taken while traveling or at funerals were subsumed under those categories.

patterns of labor allocation influenced by economic status or position within the household. Separating cane farmers from non–cane farmers (Tables 9.4 and 9.5) reveals somewhat different patterns of time use by men and women in each group, with some of that difference, not surprisingly, showing up in agriculture.

In cane-farming households, less time is spent per person per day on agricultural work than in non-cane-farming households, and women spend on average less time per day in agricultural work than do men. The finding that men in this area put more time into food-crop farming than do women differs sharply from some reported findings for other parts of Africa but is in line with other ongoing research in western Kenya (cf. Conelly, 1987).

Women in cane-farming and non-cane-farming households also dif-

Table 9.4. Women's time allocation compared for cane-farming and non-cane-farming households

Activity	Cane-farming women (hours/person/day) N = 232	Non-cane-farming women (hours/person/day) N = 126
Domestic work		
Food preparation	2.47	2.30
Food processing	0.31	0.33
Fetching water	0.52	0.45
Collecting firewood	0.22	0.16
Child care	0.75	0.57
Housecleaning-compound	0.74	0.50
Shopping	0.40	0.40
Animal care	0.10	0.04
(Subtotal)	(5.51)	(4.75)
Agricultural work		
Food-crop farming	1.51	2.06
Sugarcane farming	0.14	0.07[a]
Other farming	0.04[b]	—
Gathering greens	0.13	0.11
(Subtotal)	(1.82)	(2.24)
Other income-generating work		
Wage employment	0.04	0.02
Crafts	0.59	0.40
Marketing	0.44	0.72
Agricultural wage labor	—	0.19
(Subtotal)	(1.07)	(1.33)
Personal activities		
Sleeping	9.14	9.10
Eating	1.67	1.75
Bathing	0.65	0.68
Visiting hospital	0.19	0.16
Other	0.94	0.87
(Subtotal)	(12.59)	(12.56)
Social activities		
Resting/relaxing	1.53	1.29
Visiting with friends	0.43	0.18
Attending funerals	0.03	0.26
Community or church	0.51	0.39
Travel/transport	0.31	0.81
(Subtotal)	(2.81)	(2.93)
TOTAL	23.8[c]	23.8[c]

[a] Represents work on family's noncontracted canefields.
[b] Represents primarily labor on family's tobacco fields.
[c] Results not 24 hours because of rounding. Disaggregated categories slightly different from Table 9.1.

Table 9.5. Men's time allocation compared for cane-farming and non-cane-farming households

Activity	Cane-farming men (hours/person/day) $N=72$	Non-cane-farming men (hours/person/day) $N=74$
Domestic work		
Food preparation	0.05	0.11
Food processing	—	—
Fetching water	0.01	—
Collecting firewood	0.05	—
Child care	—	0.10
Housecleaning	—	0.02
Compound repair	0.37	0.22
Shopping	0.09	0.09
Animal care	1.87	1.78
(Subtotal)	(2.44)	(2.32)
Agricultural work		
Food-crop farming	1.59	2.99
Sugarcane farming	0.5	0.24
Other farming	0.57	0.11
Gathering greens	—	—
(Subtotal)	(2.66)	(3.34)
Other income-generating work		
Wage employment	0.71	0.68
Crafts/carpentry	0.04	0.24
Marketing	0.04	0.14
Agricultural wage labor	0.25	0.63
(Subtotal)	(1.04)	(1.69)
Personal activities		
Sleeping	8.56	8.48
Eating	1.85	1.59
Bathing	0.67	0.78
Visiting hospital	0.15	0.09
Other	0.65	1.00
(Subtotal)	(11.88)	(11.94)
Social activities		
Resting/relaxing	2.58	2.42
Visiting with friends	1.46	0.74
Attending funerals	0.05	0.18
Community or church	0.34	0.14
Travel/transport	1.41	1.07
(Subtotal)	(5.84)	(4.55)
TOTAL	23.86[a]	23.84[a]

[a] Results not 24 hours because of rounding. Disaggregated categories slightly different from Table 9.1.

fer regarding their involvement in agricultural labor. In the time-allocation study, no women from cane households reported having worked on another person's fields for pay. In contrast, women from nonsugar households reported hiring themselves out 8 percent of the total days spent in agricultural activities. Although the numbers in the time-allocation sample are small, the same trend appears in the results from a different questionnaire about hired labor. Asked about participation in hired labor over the previous week, 7.3 percent of the women from cane households reported such work, compared with 11 percent from nonsugar households. Among men, this difference is more striking. Only 2.2 percent of men in sugar households worked as hired labor, compared with 15 percent of men in nonsugar households.

Women's time use also differed in other ways. Women in cane households spend more time in domestic tasks and in craft work (primarily basket making) that can be carried out in the home while simultaneously supervising children. They also have more leisure time. In contrast, women in nonsugar households, in addition to putting slightly more time into agricultural work on food crops, spend less time on domestic work and more time in marketing and transport, as well as in hired agricultural labor. These latter activities take place away from the home and are more difficult to combine with child care, but they contribute significantly to female-controlled income. Since nonsugar households are also less likely to be polygynous, these data suggest that children in these nonsugar households might receive less of their mother's time, or possibly less time in an adult's care. In the first survey, Cogill (1987:420–21) found that children cared for by an adult did have slightly better short-term nutritional status.

Other differences appear when women are divided according to whether they have preschoolers (Tables 9.6 and 9.7). In both types of farming households, women with preschoolers spend more time in domestic work and less time in agricultural work than do their co-wives with older or no children. This finding confirms that the household division of labor is based not only on gender but also takes account of other familial responsibilities.

To summarize, men and women in the project area clearly spend their time in different ways (Table 9.8). Certain tasks are the exclusive domain of one group; other ways to pass time are given different priorities by men and women. Women are heavily occupied with domestic chores such as housekeeping, food preparation, and child care in addition to their participation in agriculture and animal care, al-

Table 9.6. Time allocation compared for women with preschoolers in cane-farming and non-cane-farming households

Activity	Cane-farming women (hours/person/day) N = 161	Non-cane farming women (hours/person/day) N = 86
Domestic work		
Food preparation	2.54	2.66
Food processing	0.33	0.43
Fetching water	0.54	0.47
Collecting firewood	0.20	0.16
Child care	0.87	0.57
Housecleaning-compound	0.75	0.58
Shopping	0.46	0.41
Animal care	0.12	0.04
(Subtotal)	(5.81)	(5.32)
Agricultural work		
Food-crop farming	1.39	2.09
Sugarcane farming	0.14	—
Other farming	0.07[a]	—
Gathering greens	0.13	0.10
(Subtotal)	(1.73)	(2.19)
Other income-generating work		
Wage employment	0.01	0.03
Crafts	0.74	0.42
Marketing	0.33	0.78
Agricultural wage labor	—	0.17
(Subtotal)	(1.08)	(1.40)
Personal activities		
Sleeping	9.20	9.10
Eating	1.63	1.79
Bathing	0.69	0.67
Visiting hospital	0.22	0.19
Other	0.88	0.72
(Subtotal)	(12.62)	(12.47)
Social activities		
Resting/relaxing	1.21	1.09
Visiting with friends	0.45	0.20
Attending funerals	0.04	0.06
Community or church	0.59	0.26
Travel/transport	0.33	0.86
(Subtotal)	(2.62)	(2.47)
TOTAL	23.9[b]	23.9[b]

[a] Represents primarily labor on family's tobacco fields.
[b] Results not 24 hours because of rounding.

Table 9.7. Time allocation compared for women with preschoolers in cane-farming and non-cane-farming households

Activity	Cane-farming women (hours/person/day) N = 71	Non-cane farming women (hours/person/day) N = 40
Domestic work		
Food preparation	2.31	1.52
Food processing	0.26	0.13
Fetching water	0.47	0.42
Collecting firewood	0.28	0.18
Child care	0.48	0.57
Housecleaning-compound	0.73	0.36
Shopping	0.25	0.38
Animal care	0.07	0.05
(Subtotal)	(4.85)	(3.61)
Agricultural work		
Food-crop farming	1.77	2.02
Sugarcane farming	0.18	0.25 [a]
Other farming	—	—
Gathering greens	0.14	0.13
(Subtotal)	(2.09)	(2.40)
Other income-generating work		
Wage employment	0.11	—
Crafts/carpentry	0.26	0.42
Marketing	0.69	0.60
Agricultural wage labor	—	0.23
(Subtotal)	(1.06)	(1.27)
Personal activities		
Sleeping	8.95	9.10
Eating	1.75	1.66
Bathing	0.59	0.69
Visiting hospital	0.14	0.10
Other	1.33	1.39
(Subtotal)	(12.76)	(12.94)
Social activities		
Resting/relaxing	2.07	1.47
Visiting with friends	0.40	0.13
Attending funerals	—	0.67
Community or church	0.34	0.67
Travel/transport	0.26	0.71
(Subtotal)	(3.07)	(3.65)
TOTAL	23.9 [b]	23.9 [b]

[a] Represents primarily labor on family's tobacco fields.
[b] Results not 24 hours because of rounding.

Table 9.8. Summary comparison of time use (hours/day) for agricultural households

			Type of activity		
	Domestic	Agriculture	Other income	Personal	Social
All men					
N = 146	2.58	3.05	1.39	11.90	5.19
All women					
N = 358	5.29	2.01	1.17	12.60	2.86
NSCF men					
N = 74	2.32	3.34	1.69	11.94	4.55
SCF men					
N = 72	2.44	2.66	1.04	11.88	5.84
NSCF women (no preschoolers)					
N = 40	3.61	2.40	1.27	12.94	3.65
SCF women (no preschoolers)					
N = 71	4.85	2.09	1.06	12.76	3.07
NSCF women (with preschoolers)					
N = 86	5.32	2.19	1.40	12.47	2.47
SCF women (with preschoolers)					
N = 161	5.81	1.73	1.08	12.62	2.62

Note: NSCF = non-sugarcane-farming; SCF = sugarcane-farming.

though the extent of their involvement in each depends in part upon their reproductive stage. Men's time is more concentrated in animal care and agricultural work; they also spend a larger part of the day in social activities and travel. Very little of their time is spent in domestic labor.

While both men and women participate in food-crop farming, they again have different responsibilities. Men's hourly contribution to agriculture is higher than expected, though consistent with observations and other studies in the area. The greater share of men's contribution to farming is in land preparation and plowing, which occupies long hours, but only for certain periods of the agricultural cycle. Women's work in farming is more heavily concentrated in weeding, harvesting, and postharvest processing and marketing.

Categorization of labor patterns by gender roles alone, however, conceals differences influenced by involvement in sugarcane production. Both men and women in cane-growing households average less time in food-crop agriculture. Both men and women in nonsugar households also hire themselves out to other farmers more than do members of sugarcane households. Women in cane households appear to have more leisure time than nonsugar women, and they also spend more of their time in house-centered domestic work. Within each farming group, men spend the greatest amount of time in agricultural work, women without preschoolers the next greatest amount of time, and women with preschoolers the least.

Significance of Findings for Policy Formulation and Technology Development

The time-allocation study provides a detailed description of labor allocation in the project area. It is clear that households that participate in sugarcane cultivation allocate household labor differently from other agricultural households, and, most important, that individuals in these households spend less time in food-crop agriculture, even though the two groups farm similar absolute areas in food crops. The question remains, however, as to how this information can be useful in formulating health and food policy initiatives or in designing new agricultural technology.

First, the time-allocation material helps to identify heterogeneity among farmers and to clarify existing labor patterns within the project area. Intercropping, a strategy currently under investigation for its pest management potential by ICIPE, is not used as much on food plots in cane-growing households and is less commonly employed on plots controlled by men. Although agricultural work is an important activity for both men and women, we expected that women's greater share of the work of planting, weeding, and harvesting would make them a more likely audience for improved intercropping practices. The time-allocation results make clear, however, that given women's already heavy work loads, intercropping is unlikely to be adopted unless it reduces their agricultural work load or greatly increases food security.

The IFPRI study concluded that participants in the sugarcane outgrowers scheme showed a definite increase in household income compared both with their preproject levels and with the levels of nonparticipants in the project. More money did not translate into significant health or nutritional benefits for preschoolers (Kennedy and Cogill, 1987:58). Neither, however, did cane growing seem to have any negative impact for preschoolers, as was first feared.

The differences in women's time use by activity group do not correlate with the lack of difference in children's nutritional status found in the IFPRI survey results. If women's time use is a critical factor in children's health and nutritional status, then greater differences in the anthropometric data and in the levels of children's caloric intake would be expected. Several interpretations are possible. Assuming the accuracy of the preschooler nutritional status measurements, it may be that women's time use is not now a major factor influencing children's nutritional status.

Other environmental factors such as water quality, sanitation, and access to medical care may play a stronger role in determining children's nutritional status. The project area suffers from one of the highest levels of child mortality and malnutrition in Kenya. Since the population in the project area is ethnically homogeneous, culturally mediated beliefs about child feeding and health standards also play an important role in defining consumption standards. These factors are equally problematic for all rural households, thus potentially negating the positive effects on household income resulting from participation in the sugar scheme.

While participation in the cane outgrowers scheme may not now be influencing children's nutritional status, it clearly has had other impacts in the project area. The time-allocation results give important insights into these changes. Documenting differences in time use related to food production and agricultural wage labor shows that the production patterns of sugar farmers and nonsugar farmers differ not only in their participation in sugarcane cropping but also in food growing. Cane cultivation ties the farmer into a complex network of labor relations both with the sugar factory and with other hired labor from within and outside the community. The income benefits from sugarcane result from these changing labor relations between farming groups and are in part supported by the labor of those not growing sugar. There are regional implications as well, since increased labor needs are currently being met by busing in workers from distant towns.

Just as labor relations between farming households are affected, so are gender relations regarding food production within the household. The move into cane farming has decreased the time men put into food production. Women in cane-farming households also show a lower time input into food production than women in non-sugarcane-farming households, but their responsibility for food plots is greater. Greater use of hired labor in cane households parallels these shifts.

While the sugarcane scheme has resulted in higher aggregate household incomes for its participants, nutritional status has not measurably improved (Kennedy and Cogill, 1987; Kennedy, 1988), and it is not clear that women-controlled income has increased. Men and women, and adults and children, are being affected in different ways. The changes in the allocation of time across and within agricultural households suggest that important shifts in productive behaviors are occurring in areas other than nutrition that have potentially less favorable impacts on the welfare of the project area population.

Time-allocation studies are useful ways of looking at nuanced differences in the social organization of labor within households, between households, and across the community. They can corroborate or question other findings about labor patterns. The intriguing differences in time use associated with social roles within and between households in this study point to other research questions about the relationships between changing production processes and the value of time.

Bibliography

Berry, Sara. "Macro-Policy Implications of Research on Rural Households and Farming Systems." In Joyce L. Moock, ed., *Understanding Africa's Rural Households and Farming Systems*. Boulder, Colo.: Westview Press, 1986.

Bryson, Judy C. "Women and Agriculture in Sub-Saharan Africa: Implications for Development (An Exploratory Study)." In N. Nelson, ed., *African Women in the Development Process*. London: Frank Cass, 1981.

Chaiken, Miriam. "Participatory Development and African Women: A Case Study from Western Kenya." In Miriam S. Chaiken and Anne K. Fleuret, eds., *Social Change and Applied Anthropology: Essays in Honor of David W. Brokensha*. Boulder, Colo.: Westview Press, 1990.

Cogill, Bruce. "The Effects on Income, Health, and Nutritional Status of Increasing Agricultural Commercialization in South-West Kenya." Ph.D. diss., Cornell University, Ithaca, N.Y., 1987.

Conelly, William T., et al. "Household Labor Allocation in an Intensive Crop/Livestock Farming System in Western Kenya." Paper presented at the Farming Systems Research Symposium, University of Arkansas, Fayetteville, 1987.

Davison, Jean, ed. *Agriculture, Women, and Land*. Boulder, Colo.: Westview Press, 1988.

Dey, Jennie. "Gambian Women: Unequal Partners in Rice Development Projects?" In N. Nelson, ed., *African Women in the Development Process*. London: Frank Cass, 1981.

Gross, David. "Time Allocation: A Tool for the Study of Cultural Behavior." *Annual Review of Anthropology*, 13:519–58, 1984.

Guyer, Jane. "Dynamic Approaches to Domestic Budgeting: Cases and Methods from Africa." In Daisy Dwyer and Judith Bruce, eds., *A Home Divided: Women and Income in the Third World.* Stanford, Calif.: Stanford University Press, 1988.

Hay, Margaret Jean, and Sharon Stichter, eds. *African Women South of the Sahara.* London: Longman, 1984.

Jones, Christine W. "Intra-household Bargaining in Response to the Introduction of New Crops: A Case Study from North Cameroon." In J. L. Moock, ed., *Understanding Africa's Rural Households and Farming Systems.* Boulder, Colo.: Westview Press, 1986.

Kennedy, Eileen T. "The Effects of Sugarcane Production on Food Security, Health, and Nutrition in Kenya: A Longitudinal Analysis." Research Report 78. Washington, D.C.: International Food Policy Research Institute, 1989.

Kennedy, Eileen T., and Bruce Cogill. "Income and Nutritional Effects on the Commercialization of Agriculture in Southwestern Kenya." Research Report 63. Washington, D.C.: International Food Policy Research Institute, 1987.

Little, Peter. "Institutional Dynamics and Development in the Tana Basin, Kenya." In Miriam S. Chaiken and Anne K. Fleuret, eds., *Social Change and Applied Anthropology: Essays in Honor of David W. Brokensha.* Boulder, Colo.: Westview Press, 1990.

Moock, Joyce L. ed. *Understanding Africa's Rural Households and Farming Systems.* Boulder, Colo.: Westview Press, 1986.

Nelson, Nici, ed. *African Women in the Development Process.* London: Frank Cass, 1981.

Rubin, Deborah. "Changing Practices in a Sugarcane Growing Community." Final Report prepared for U.S. AID, Office of Policy and Program Coordination (July). Washington, D.C.: International Food Policy Research Institute, 1988.

von Braun, Joachim, and Eileen T. Kennedy. "Commercialization of Subsistence Agriculture: Income and Nutritional Effects in Developing Countries." Working Papers on Commercialization of Agriculture and Nutrition, no. 1. Washington, D.C.: International Food Policy Research Institute, 1986.

Watts, Michael. "Idioms of Land and Labor: Producing Politics and Rice in Senegambia." Paper presented at the Popular Culture and Popular Protest in Africa conference, Stanford University, 1989.

Webb, Patrick. "Intrahousehold Decision-making and Resource Control: The Effects of Rice Commercialization in West Africa." Working Papers on Commercialization of Agriculture and Nutrition, no 3. Washington, D.C.: International Food Policy Research Institute, 1989.

Williams, Simon. "The Mumias Sugar Company: A Nuclear Estate in Kenya." In S. Williams and R. Karen, eds., *Agribusiness and the Small Scale Farmer.* Boulder, Colo.: Westview Press, 1985.

10

Comparative Advantage of Crop-Livestock
Production Systems in Southwestern Nigeria
and the Technical Research Implications

Timothy Olalekan Williams

Introduction

In southwestern Nigeria, crop and livestock production systems have
traditionally been operated as separate farm enterprises. Crop production, the major farming enterprise, is largely undertaken within
the bush fallow or shifting-cultivation system. Rapid changes in rural
population and increasing pressure on agricultural land have resulted
in shorter fallow periods, thus making the maintenance of crop yields
and soil fertility a major problem. The decreasing importance of the
bush fallow system as a strategy for maintaining soil fertility has led
to the development and subsequent introduction to farmers of several
innovations that not only maintain soil fertility under continuous
cropping but also encourage increased integration of crop and livestock production systems. Examples of such innovations include the
techniques of alley cropping and alley farming.

While it appears that there are advantages to be derived from a
mixed farming system that links crop and livestock production, the
relative efficiency of resource use among competing crop and livestock enterprises still needs to be determined to demonstrate the viability of the alternative activities to potential investors and to help
guide future investment in research. This chapter considers the com-

parative advantage and policy incentives for maize–small ruminant production systems in southwestern Nigeria, using the resource-cost ratio (RCR) approach to analyze the competitiveness and real cost to Nigeria of alternative maize–small ruminant production enterprises.[1]

This approach differs from previous economic analyses of alley cropping and alley farming in that it weighs, from both private and social perspectives, the profitability of alternative crop-livestock enterprises in southwestern Nigeria. The method will help farmers and policymakers to decide the relative emphasis to be given to crop and mixed farming activities. It should also provide biological scientists with a more accurate estimate of productivity gains needed to make integrated crop and livestock enterprises competitive within the farming systems of southwestern Nigeria.

The Study Area and Alley-Farming Systems

Southwestern Nigeria lies within the humid zone. Annual rainfall, which is bimodally distributed, ranges from 1000 to 2000 mm. Mean temperatures (27–30° C) and relative humidity (80–90 percent) are high. Agriculture in this area is predominantly based on shifting cultivation or bush fallow rotation. A cropping period of three to four years is followed by an extended fallow period to restore soil fertility (Jaiyebo and Moore, 1964). Landholdings are small (1–3 ha), and land is hoe cultivated. Arable crops are cereals, roots, and vegetables, of which maize, cassava, and yam are the most widely grown for home consumption and cash sale.

The dominant ruminant livestock species in this area are trypanotolerant West African dwarf sheep and goats. Although most farm households keep sheep and goats, small ruminant keeping is generally not well integrated with crop production in that few forage crops are grown and manure is rarely applied to crops. The animals have traditionally been kept exclusively for meat, and they are reared more for sale than for home consumption. The average flock size per owner

---

[1] The resource-cost ratio is a measure of the total cost of production when prices are adjusted for taxes and subsidies and resources are valued in alternative uses. It is a measure of relative economic efficiency and is calculated as the ratio of domestic factor costs valued at world price equivalent to value added at world prices. A RCR greater than 1 implies that the value of domestic resources used in producing an output is greater than the value of the foreign exchange saved or earned, and vice versa for a RCR less than 1. For a more detailed discussion of the methodology, see Page and Stryker, 1981; Byerlee and Longmire, 1986.

is three or four animals, with goats predominating. Goats are generally more popular than sheep, possibly because they are more prolific, are easier to manage, and require less initial capital outlay (Mack, 1983). Flock size and productivity in both species are limited mainly by disease, feed availability, and poor management.

While small ruminants presently play a relatively minor role in the farming systems of this area, the potential exists for improved integration of crop and livestock production systems through the cultivation of leguminous browse trees within arable crop farms. The realization of this potential has been a major research preoccupation of the International Livestock Centre for Africa's (ILCA) Humid Zone Program based in Ibadan, Nigeria. The program has promoted the technique of alley farming as a means of improving livestock production while also enhancing arable crop production.

Alley farming entails planting leguminous trees in hedgerows 4 m apart with arable crops grown between the rows. The trees are pruned regularly and a portion of the foliage is used as cut-and-carry feed for small ruminants, while the remaining foliage is utilized as green manure for soil fertility maintenance. The system thus differs from the alley-cropping technique considered in this chapter, in which the all of the foliage is applied to the soil as mulch.

Experimental work on-station and trials on farmers' fields have demonstrated that in comparison with traditional farming systems, alley farming offers increased crop yields and provides high-quality fodder for animals (Atta-Krah et al., 1986; Reynolds and Adeoye, 1986). A number of economic analyses have also shown the innovation to be financially profitable for farmers (Ngambeki, 1985; Sumberg et al., 1987), but none of these have considered the *social* profitability of alley farming. Given that private profitability and social profitability differ as a result of government policy interventions, it is worthwhile to consider briefly the impact of government price, trade, and exchange rate policies on crop-livestock production activities in Nigeria.

The Policy Environment for Crop and Livestock Production

The dominant feature of the Nigerian policy environment from 1973 to 1983 was the increase in oil export revenues. The oil boom and the related management of the exchange rate had a number of effects on agriculture. Expenditures of oil revenues, particularly on urban construction activities, lured workers away from agriculture and raised the wages of those who remained. Between 1973 and 1980 the real exchange rate appreciated by 61 percent, partly as a result of massive capital inflows associated with the oil boom and partly because of the government's failure to depreciate the naira to reflect Nigeria's relatively high inflation rate (Oyejide, 1986). The overvalued exchange rate was sustained by periodic import restrictions and exchange control regulations. Short-term variations in quantitative import restrictions caused substantial price instability for several crop and livestock products during this period.

Regarding the specific products considered in this chapter, maize was the only item that came under direct government agricultural pricing policy intervention. Between 1976 and 1986, guaranteed minimum prices (GMPs) were established for maize and other cereals. The GMPs served as a below-market safety net rather than as a floor, in that they were deliberately set too low. The prices were often set at fixed levels for several years at a time. The result was that in all cases, the GMPs were below farm-gate prices and largely inoperative. Over the same period livestock prices were determined by market forces. Figure 10.1 shows that farm-gate prices for maize were fairly steady in real terms between 1973 and 1986. Massive maize imports over most of this period served to depress the prices received by farmers through the appreciation of the real exchange rate, which made imports cheaper than domestically produced maize. Farm-gate prices for goats and sheep, on the other hand, rose sharply in real terms between 1973 and 1976. The drought in northern Nigeria during 1972–74 partly accounted for this rise. Real prices for these animals declined between 1976 and 1983, but they appear to have risen again since 1984.

The overvalued naira, quantitative import restrictions, and increasing labor costs between 1973 and 1984 were the main reasons behind the upward trend in nominal protection coefficients (NPCs) for maize, goats, and sheep, as shown in Figure 10.2.[2] The NPCs have been estimated using the official exchange rate. Figure 10.2 indicates that although domestic maize prices were below import parity prices between 1973 and 1975, all three products received protection from external competition for most of the period under consideration. The level of protection for sheep and goats was higher than the protection

[2] The nominal protection coefficient measures the extent to which domestic prices diverge from world parity prices.

Figure 10.1. Real farm-gate prices of maize, goats, and sheep in Nigeria, 1973–86

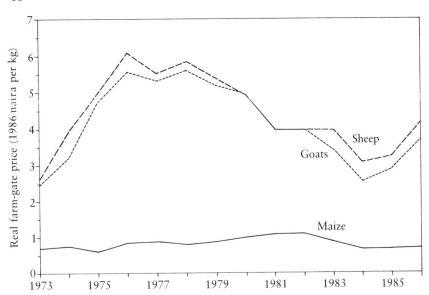

Source: FAO (1984) and Central Bank of Nigeria (1986).

given to maize between 1975 and 1979, but the position has since been reversed due to the declining world price equivalent for maize. The upward trend in NPCs for maize, sheep, and goats is dampened when the exchange rate is adjusted to account for the overvaluation of the naira.[3]

Government policies have also affected agricultural inputs. Most of the special programs initiated during the 1970s to boost food production used farm input subsidies as their major incentive (Idachaba et al., 1980). Such subsidies continued to be an important component of Nigeria's agricultural development strategy until 1986, when the structural adjustment program commenced. Subsidies extended over a wide range of inputs, such as fertilizer, improved seeds, and pesti-

[3] The adjusted exchange rate is meant to correct for distortions in the official exchange rate. The extent of overvaluation of the latter was estimated using the differential inflation rate between domestic prices (approximated by the consumer price index) and foreign prices (based on the consumer price index of industrialized countries). The year Nigeria's currency was changed from the pound to the naira (1973) was used as the base year. This method of correcting exchange rates, however, provides only a rough estimate of the extent of overvaluation because it ignores factors other than inflation differential (e.g., high tariff protection and import controls) that lead to exchange rate distortions.

Figure 10.2. Nominal protection coefficients (NPCs) for maize, goats, and sheep in Nigeria, 1973–86

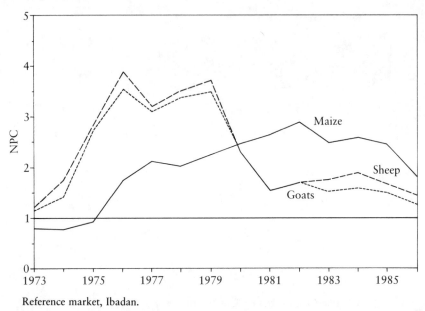

Reference market, Ibadan.

cides, and on a wide variety of services, such as land-clearing and irrigation, as well as on various credit schemes. Fuel costs remain subsidized. Most smallholders, however, did not benefit from these subsidy schemes due to preferential treatment accorded to larger farmers (Idachaba et al., 1980). Labor, the most important factor input into crop and livestock production in Nigeria, was not subsidized.

What seems clear from this brief analysis is that the overall effect of government policies has been to raise the domestic prices of maize, goats, and sheep above their equivalent world prices, resulting in varying degrees of protection for domestic production. Until 1985, domestic prices for maize, goats, and sheep were simultaneously high by international standards and low in terms of being able to attract domestic resources to increase production.

This view is reinforced by the fact that imports of these commodities remained quite high until about 1985. This picture, which largely depicts the situation in the agricultural sector as a whole, is in sharp contrast to the enhanced profitability of investments in the construction and service sectors during the same period (World Bank, 1984). The next section provides the technical parameters needed for the private and social profitability calculations of alternative maize and small ruminant enterprises in southwestern Nigeria.

Alternatives Examined and Data Sources

Four enterprises were considered for the profitability calculations: maize grown using the traditional method of production, alley cropping with maize, alley farming with goats, and alley farming with sheep. The last three enterprises reflect the kind of farming activities being promoted in southwestern Nigeria by ILCA and the International Institute of Tropical Agriculture (IITA).

The alley-cropping method involves planting maize between rows of already established *Leucaena leucocephala* trees. The trees are pruned regularly and the entire foliage is applied to the soil as mulch. It is assumed that on-farm, the trees produce 3000 kg/ha of foliage dry matter per year. At a nitrogen content of 3.5 percent, this foliage contains 105 kg N/ha per year. Research results at IITA indicate that the marginal physical product of 1 kg mulch-N varies between 5 and 10 kg of maize depending on whether the mulch is applied to the soil surface or incorporated into the soil (Kang et al., 1981). The base maize yield is taken to be 1 metric ton/ha, with an annual decline of 10 percent. In the first year, the entire tree foliage applied as mulch will increase maize yields above the base level. In subsequent years, part of the mulch will be needed to maintain maize yields at the base level, while the remaining mulch will serve to increase maize yields above the base.

The alley-farming enterprises with goats and sheep are based on a cut-and-carry feeding system.[4] It is assumed that 75 percent of the leucaena tree foliage is returned to the soil as mulch to maintain and increase maize yields as described above, while the remaining foliage is used as supplementary feed for breeding animals. Daily feed intake is set equal to 4 percent of the animal's live weight. The base rate of supplementation—tree foliage intake—is taken to be equal to 25 percent of the total daily feed intake.

Technical parameters were employed as described in Tables 10.1

---

[4] While alley-farming enterprises that involve direct grazing of trees and natural regrowth by animals eliminate the problem of manure loss and permit full exploitation of the nutrient-recycling capabilities of livestock, they are not considered here, principally because the dominant position of tree and food-crop production in the area being studied makes investment in small ruminants to the exclusion of crop-farming activities highly unlikely.

Table 10.1. Technical parameters for the production of rainfed maize under traditional and alley-cropping systems in southwestern Nigeria, 1984

	Maize traditional system	Alley cropping with maize[a]
Inputs		
Labor (man-days/ha)		
Land preparation	16	24
Planting	5	7
Weeding	25	23
Pruning/mulching	—	12
Harvesting	16	24
TOTAL LABOR	62	90
Seed (kg/ha)	20	20
Alley trees foliage (kg-DM/ha)[b]	—	3000
Output (metric tons/ha)		
Maize[c]	1.00	1.40

Note: It is assumed that under both cropping systems, one crop of maize is grown per year without chemical fertilizer.

[a] Maize planted between rows of already established leucaena trees. Labor inputs required for the establishment of the trees are included in the labor-use figures shown here.

[b] At a nitrogen content of 3.5 percent, this foliage contains 105 kg N. It is assumed that mulch is left on the soil surface and its marginal physical product is 5 kg maize per kg mulch-N; DM = dry matter.

[c] The maize yield increment under alley cropping is less than the expected 500 kg because trees take up some land space and this results in a slightly reduced maize density.

and 10.2. These parameters were unchanged from year to year. Parameters for maize grown traditionally and alley cropping with maize were obtained from IITA annual reports (IITA, 1981–84) and Ngambeki (1985). These data were supplemented with information collected from a farm management survey conducted by the author in three villages in southwestern Nigeria from March to September 1987. Animal production parameters were taken from village studies conducted by ILCA in southwestern Nigeria (Mack, 1983; Sumberg, 1985; Upton, 1985). Price data were obtained from various sources. Market prices were taken from the FAO (1984), Okali and Upton (1985), the Central Bank of Nigeria (1986), and the Federal Livestock Department (unpublished data).

The world price equivalents for maize, sheep, and goats are long-run expected prices adjusted for domestic transport and distribution costs. The world price for maize was taken from Duncan (1984). Data obtained from FAO trade yearbooks for imports of live small ruminants into Nigeria from Niger were used to estimate a long-term trend

Table 10.2. Technical parameters for maize–small ruminant alley-farming systems in southwestern Nigeria, 1984

	Goats	Sheep
Inputs		
Labor (man-days/ha)		
Land preparation	24	24
Planting	7	7
Weeding	23	23
Pruning/mulching	10	10
Harvesting	23	23
Feeding and animal care	10	10
TOTAL LABOR	97	97
Seed (kg/ha)	20	20
Alley trees foliage (kg-DM/ha)[a]	3000	3000
Breeding animals[b]		
Goats (17 kg LW)	12	—
Sheep (25 kg LW)	—	8
Outputs		
Maize (metric tons/ha)	1.3	1.3
Net animal output (dam/yr)[c]	19.40	49.40

[a] About 75 percent of the available foliage is utilized as mulch, while the remaining 25 percent is used as fodder. It is assumed that mulch is left on the soil surface and its marginal physical product is 5 kg maize per kg mulch-N.

[b] For both goats and sheep, daily dry matter feed intake is assumed equal to 4 percent of the animal's live weight (LW). Foliage intake is equal to 25 percent of the total daily feed intake.

[c] Net animal output depends on a number of factors, including parturition interval, litter size, live weight (at 12 months), price, and mortality rate. Sheep yield substantially higher outputs than goats because sheep are heavier, have lower mortalities, and are higher priced (Upton, 1985).

line from which the import prices for sheep and goats were calculated.[5]

Private and Social Profitability Calculations

Farm budgets were constructed for each major enterprise as a mea-

[5] Import prices for sheep were estimated from the long-term trend line, $\ln P = -1.608 + 0.1195\ T$, for the period 1968–86, based on the f.o.b. prices of sheep from the Niger republic as given in the FAO trade yearbooks. (T = time; i.e., a given year.) International prices for goats were unavailable, but based on the knowledge that prices of sheep and goats move in the same direction—though goat prices are usually lower—import prices obtained from the above trend line were discounted by 15–20 percent to obtain import prices for goats. The discount reflects the price differential that prevailed domestically between sheep and goat prices during the period of analysis.

Table 10.3. Budgets for maize–small ruminant enterprises in southwestern Nigeria, 1984

	Maize grown under traditional practice	Alley cropping with maize	Alley farming[a] Goats	Alley farming[a] Sheep
Labor costs (per ha)				
Land preparation	80.00	120.00	120.00	120.00
Other farm operations	230.00	330.00	365.00	365.00
Seed	16.00	16.00	16.00	16.00
Annual tool cost	16.08	16.08	20.11	20.11
Annual cost of investment in animals	—	—	137.77	161.45
Interest on working capital	4.32	6.15	6.44	6.44
TOTAL COST	346.40	488.23	665.32	689.00
TOTAL REVENUE	600.00	840.00	1027.93	1230.53
NET PROFIT	253.60	351.77	362.61	541.53

Note: Costs and revenues are given in naira.

[a] Budgets for alley-farming enterprises were constructed under the assumption that 75 percent of tree foliage is utilized as mulch while the remaining 25 percent is used as fodder for animals. The fodder intake is assumed to lead to 6.5 percent increase in productivity in the case of goats and 14 percent increase in the case of sheep (ILCA, 1988).

sure of private profitability and also as a basis for calculating resource cost ratios (Table 10.3). The budgets were estimated using farmer prices, and they indicate the private profitability of the different enterprises considered.

Labor was the single most important cost item, accounting for 70–90 percent of total private costs in 1984. The annual user cost (depreciation plus interest) of hand tools was a small proportion of the total cost. The annual cost of investment in animals, derived by using an annuity factor to adjust the initial cost of animals, accounted for 21–23 percent of the total private cost of the alley-farming systems.[6]

The cost of working capital was calculated using the market interest rate of 10 percent. Working capital is tied up in seeds and land preparation labor for about three months because these items are invested in at the start of the growing season. For labor used in other farm operations, working capital is tied up for approximately one month.

On the basis of private profitability calculations, Table 10.3 indicates that alley farming with sheep was the most profitable enterprise in 1984, followed by alley farming with goats. Alley farming with

[6] An annual capital recovery factor was used to adjust the initial cost of breeding animals in order to obtain the annual cost of investment in animals. This annuity includes both interest and depreciation on the animals.

Table 10.4. Calculation of resource-cost ratios for maize–small ruminant enterprises in southwestern Nigeria, 1984

	Maize grown under traditional practice	Alley cropping with maize	Alley farming	
			Goats	Sheep
Inputs (per ha)				
Tradables				
Seed	9.40	9.40	9.40	9.40
Fuel	2.82	3.95	6.37	5.47
Nontradables				
Labor	310.00	450.00	485.00	485.00
Costs of tools	20.49	20.49	25.62	25.62
Cost of investment in animals	—	—	229.06	272.16
Marketing & distribution costs	6.58	9.21	14.85	12.75
Interest	10.38	14.97	15.69	15.69
Land-opportunity costs[a]	149.98	110.33	149.98	149.98
Outputs				
Tradables				
Maize	470.00	658.00	611.00	611.00
Goats	—	—	292.41	—
Sheep	—	—	—	538.81
Social profitability	−39.65	39.65	−32.56	173.74
Opportunity cost of domestic resources	497.43	605.00	920.20	961.20
Value added (tradables)	457.78	644.65	887.64	1134.94
RCR	1.09	0.94	1.04	0.85

Note: Costs and revenues are given in naira.

[a] Residual return to land in best competing alternative valued at world price equivalent; i.e., maize grown under traditional methods in the case of alley cropping and alley cropping in the case of maize grown traditionally and under alley-farming systems.

goats was less profitable than alley farming with sheep because of the lower prices per kilogram live weight of goats when compared with sheep. Alley cropping with maize was almost as profitable as alley farming with goats because the marginal value product of mulch-nitrogen in alley cropping with maize was higher than the marginal value product of tree foliage–nitrogen fed to goats.

Table 10.4 shows the social profitability and RCR calculations for the enterprises examined. Outputs and inputs have been divided into tradable and nontradable components. As explained earlier, import prices for maize, goats, and sheep were based on expected long-run prices. These prices were adjusted for transport and distribution costs.

Owing to the existence of an active rural labor market with mini-

Table 10.5. Resource-cost ratios by enterprise, southwestern Nigeria, 1984–86

	Maize traditional system	Alley cropping with maize	Alley farming	
			Goats	Sheep
1984	1.09	0.94	1.04	0.85
1985	1.25	0.83	1.02	0.83
1986	1.31	0.80	1.05	0.86

mum distortion of wages, the market wage rate was taken to reflect the opportunity cost of labor. For the cost of capital, a real rate of interest of 25 percent was used. This rate represents an educated guess based on the interest rates charged by commercial banks on nonagricultural loans and the rates charged by private moneylenders in rural areas.

In the absence of reliable data, the costs of marketing the final output were simply divided into fuel (tradable) and domestic labor and capital (nontradable) and valued accordingly. Returns to land were calculated as a residual in the best competing alternative to represent the case where land is limiting.

The results in Table 10.4 indicate that net social profitability is lower than net private profitability for all enterprises examined, showing that government policies subsidized maize and small ruminant production in 1984. Producers in this instance earned higher profits than they would have in the absence of such policies. This result is not surprising given the estimates of NPCs and the description of the policy environment in Nigeria presented earlier.

The RCR for alley farming with sheep in 1984 was less than 1, while the ratio for alley farming with goats was slightly greater than 1. The results in Table 10.5 were derived by assuming the same annual percentage decline in maize yields for all enterprises and reducing harvesting labor accordingly. The results confirm the pattern shown in Table 10.4. The RCRs for alley farming with sheep were consistently less than 1, while the ratios for alley farming with goats remained consistently above 1.

Sensitivity Analysis

The empirical results discussed above are based on certain assumptions concerning the method of mulch application and the rate at which

Table 10.6. Resource-cost ratios for maize–small ruminant enterprises under various assumptions, southwestern Nigeria, 1984–86

	Maize traditional system	Alley cropping with maize	Alley farming Goats	Alley farming Sheep
RCRs with mulch incorporation				
1984	1.52	0.73	1.07	0.90
1985	1.74	0.64	1.10	0.92
1986	1.86	0.60	1.14	0.95
RCRs with 50% supplementation rate for animals				
1984	1.09	0.94	1.05	0.90
1985	1.25	0.83	1.02	0.87
1986	1.31	0.80	1.04	0.89

animals are supplemented with tree foliage. It is interesting to examine the effect of variations in these assumptions on the empirical results.

The upper part of Table 10.6 shows the effect on the RCRs of incorporating mulch into the soil. The resulting increases in the yield of maize under the alley systems and the corresponding additional labor needed for harvesting have been taken into consideration. Maize cultivated by the traditional method becomes clearly uncompetitive when compared with the other enterprises. With mulch incorporation, all other things being equal, productivity gains of about 40–45 percent per animal are needed to make alley farming with goats competitive with alley cropping with maize. When compared with the results in Table 10.5, alley farming with sheep appears to be moderately sensitive to the method of mulch application, but alley cropping with maize appears to have a comparative advantage over all the other enterprises in this case.

Raising the tree foliage intake (supplementation rate) to 50 percent of the total daily feed intake leads to about a 28 percent increase in net output per ewe and a 13 percent increase in net output per doe (ILCA, 1988). Compared with the base rate of supplementation, the number of animals that could be supported on the available foliage is reduced by half because the amount of foliage is fixed. Labor input for animal feeding and care are also reduced. Based on these assumptions, the lower part of Table 10.6 indicates that alley farming with goats does not have a comparative advantage over alley cropping with maize. Alley farming with sheep, however, appears to be competitive when compared with alley cropping with maize.

Generally, one can conclude that the estimates of RCRs for the various enterprises are moderately sensitive to maize yields and net animal output. Alley farming with sheep appears likely to maintain its competitiveness in the medium term.

Conclusions

This chapter has presented an analysis of three new systems of crop and livestock production currently being promoted by IITA and ILCA in southwestern Nigeria. A major objective of this study has been to determine the relative efficiency of resource use among competing crop and livestock activities. Based on the analysis of comparative advantage presented in this paper, there is a fairly strong case for promotion of research and extension in alley-cropping and alley-farming methods in southwestern Nigeria. There is a comparative advantage in alley cropping with maize relative to the traditional method of growing maize and alley farming with goats.

It appears that considerable productivity gains need to be achieved to make alley farming with goats a competitive enterprise in southwestern Nigeria. Meeting this challenge in the long run would be particularly relevant to farmers in the study area, given that goats are more prolific and more popular than sheep. However, within the short- to medium-term framework, research attention should focus on alley farming with sheep, as it appears to give a higher national return to land than alley farming with goats.

Bibliography

Atta-Krah, A. N., J. E. Sumberg, and L. Reynolds. "Leguminous Fodder Trees in The Farming System—An Overview of Research at the Humid Zone Programme of ILCA in Southwestern Nigeria." In I. Haque, S. Jutzi, and P. J. Neate, eds., *Potentials of Forage Legumes in Farming Systems of Sub-Saharan Africa*. Proceedings of a Workshop held at ILCA, Addis Ababa, Ethiopia, September 16–19, 1985. Addis Ababa, Ethiopia: ILCA, 1986.

Byerlee, D., and J. Longmire. "Comparative Advantage and Policy Incentives for Wheat Production in Rainfed and Irrigated Areas of Mexico." CIMMYT Economics Program, Working Paper no. 01/86, 1986.

Central Bank of Nigeria. *Annual Report and Statement of Accounts*. Lagos, Nigeria: Government Printer, 1986.

Duncan, R. C., ed. "The Outlook for Primary Commodities, 1984 to 1995."

World Bank Staff Commodity Working Paper no. 11. Washington, D.C.: World Bank, 1984.

FAO (Food and Agriculture Organization). *Statistics on Prices Received by Farmers*. Rome: FAO, 1984.

Federal Livestock Department. *Nigerian Livestock Information Service Annual Bulletin* (various issues).

Idachaba, F. S., et al. *The Green Revolution: A Food Production Plan for Nigeria*. Lagos, Nigeria: Federal Ministry of Agriculture, 1980.

IITA (International Institute for Tropical Agriculture). *Annual Reports, 1981–84*. Ibadan, Nigeria.

ILCA (International Livestock Centre for Africa). *Annual Report, 1987*. Addis Ababa, Ethiopia: ILCA, 1988.

International Monetary Fund. *International Financial Statistics* Yearbook. Geneva: IMF, 1987.

Jaiyebo, E. O., and A. W. Moore. "Soil Fertility and Nutrient Storage in Different Soil Vegetation Systems in Tropical Rain Forest Environment." *Tropical Agriculture*, (Trinidad) 41:129–30, 1964.

Kang, B. T., and L. Reynolds. *Alley Farming in the Humid and Subhumid Tropics*. Proceedings of the Alley Farming Workshop, Ibadan, Nigeria, March 10–14, 1986. Ibadan, Nigeria: IITA and ILCA, 1986.

Kang, B. T., et al. "*Leucaena* (*Leucaena leucocephala* [Lam] de wit) Prunings as Nitrogen Source for Maize (*Zea mays* L.)." *Fertilizer Research* 2:279–87, 1981.

Mack, S. D. *Evaluation of the Productivities of West African Dwarf Sheep and Goats*. Humid Zone Program, Document 7. Addis Ababa, Ethiopia: ILCA, 1983.

Ngambeki, D. S. "Economic Evaluation of Alley Cropping Leucaena with Maize-Maize and Maize-Cowpea in Southern Nigeria." *Agricultural Systems*, 17:243–58, 1985.

Okali, C., and M. Upton. "The Market Potential for Increased Small Ruminant Production in Southwest Nigeria." In J. E. Sumberg and K. Cassaday, eds., *Sheep and Goats in Humid West Africa*. Proceedings of the Workshop on Small Ruminant Production Systems in the Humid Zone of West Africa, Ibadan, Nigeria, January 23–26, 1984. Addis Ababa, Ethiopia: ILCA, 1985.

Oyejide, T. A. *The Effects of Trade and Exchange Rate Policies on Agriculture in Nigeria*. Research Report 55. Washington, D.C.: IFPRI, 1986.

Page, J. M., Jr., and J. D. Stryker. "Methodology for Estimating Comparative Costs and Incentives." In Scott R. Pearson et al., eds., *Rice in West Africa: Policy and Economics*. Stanford, Calif.: Stanford University Press, 1981.

Reynolds, L., and S. A. O. Adeoye. "Alley Farming with Livestock." Paper presented at the International Workshop on Alley Farming for Humid and Subhumid Regions of Tropical Africa, Ibadan, Nigeria, March 10–14, 1986. Addis Ababa, Ethiopia: ILCA, 1986.

Sumberg, J. E. "Small Ruminant Feed Production in a Farming Systems Context." In J. E. Sumberg and H. Cassaday, eds., *Sheep and Goats in Humid West Africa*. Proceedings of the Workshop on Small Ruminant Production Systems in the Humid Zone of West Africa, Ibadan, Nigeria, January 23–26, 1984. Addis Ababa, Ethiopia: ILCA, 1985.

Sumberg, J. E., et al. "Economics of Alley Farming." *ILCA Bulletin*, 28:1–6, 1987.

Upton, M. "Models of Improved Production Systems for Small Ruminants." In J. E. Sumberg and H. Cassaday, eds., *Sheep and Goats in Humid West Africa.* Proceedings of the Workshop on Small Ruminant Production Systems in the Humid Zone of West Africa, Ibadan, Nigeria, January 23–26, 1984. Addis Ababa, Ethiopia: ILCA, 1985.

World Bank. *Nigeria: Agricultural Sector Memorandum.* Volume 2, Main Report. Washington D.C.: World Bank, 1984.

11

Sense and Sustainability: Sustainability as an Objective in International Agricultural Research

John K. Lynam and Robert W. Herdt

Introduction

The international agricultural research establishment, having achieved a degree of visibility following the so-called Green Revolution, has attracted the attention of international gadflies and social critics, with each new generation proclaiming new criteria for the evaluation of agricultural technology. Previous criteria have included production, technology for small farmers, welfare of low-income consumers, technology for women, diversification, and stability. Sustainability is the latest twist in the continuing elaboration of criteria by which agricultural development is defined and agricultural technology is evaluated.[1]

A spate of recent publications has explored the possible implications for the environment, human welfare, and the world food balance signaled by indications that the agricultural resource base in the

This chapter is a modified version of our article "Sense and Sustainability: Sustainability as an Objective in International Agricultural Research," *Agricultural Economics*, 3:381–98, 1989.
[1] Two facts illustrate the gadfly nature of the word *sustainability:* One is hard-pressed in 1992 to find an agricultural conference, publication, or new research program that does not include *sustainability* in its title, but the word does not appear in *Webster's New Collegiate Dictionary* or in the dictionary of WordPerfect 5.0.

tropics is being overexploited. These reflect a concern for the future and at their most profound represent an imperative for the world to plan more thoughtfully, in some instances to reconsider, the progress of tropical agricultural development. While we fully accept the need to be concerned about future development paths, in this chapter we attempt the more limited and mundane task of considering how sustainability concerns might be incorporated into the research activities of the international agricultural research centers.

The sustainability criterion has emerged from the recent visibility of ecologists in agricultural development in the Third World. Nevertheless, the theme of sustainability has been appropriated by a range of institutions interested in agricultural development and now includes a broad array of concerns about the maintenance of the resource base to ensure future levels of agricultural production. These concerns encompass such diverse areas as loss of genetic diversity in crop species, tropical deforestation, soil erosion, the effect of agrochemicals on the environment, and the implications of global warming for agricultural production. Those concerns are high on the agenda of the international agricultural research establishment; however, because sustainability is essentially a set of concerns about future conditions, it is not easy to translate these into operational agricultural research activities.

The sustainability concept could be incorporated into the research process at three levels: (1) as an evaluation criterion in technology testing, (2) as a design criterion in the creation of crop technologies, and (3) as a set of concerns (objectives?) around which to organize research. Progression from (1) to (3) signifies the upgrading of sustainability from an intermediate to an end objective (Pinstrup-Andersen and Franklin, 1977). Moreover, at the first level, sustainability could be incorporated into existing research programs, while the third level implies the reorganization of the research process.

Sustainability as an Evaluation Criterion

If sustainability is to be used as an evaluation criterion, a precise and unambiguous definition is necessary. No one would disagree with the statement that implementing sustainable development would "ensure that humanity meets the needs of the present without compromising the ability of future generations to meet their own needs" (*Our Common Future,* report of the World Commission on Environment and Development). But this does not provide a criterion for agricultural researchers. Papers by the Technical Advisory Committee (TAC) of the Consultative Group on International Agricultural Research (CGIAR, 1988) and the World Resources Institute (Dover and Talbot, 1987) provide not so much a definition as a characterization of the term. The TAC suggests that it is not stability. Dover and Talbot appear to suggest that stability, used in an ecological way, comes closest to defining sustainability. Conway (1985) offers a precise definition, but in Dover and Talbot's view his is merely the definition for resilience, another specific ecological concept. Holling (1973) provides a discussion of resilience and stability in ecological theory. All this points to a concept entailing substantial ambiguity in any particular application.

Conway's (1985) definition nevertheless provides a useful starting point; namely, "sustainability is the ability of a system to maintain productivity in spite of a major disturbance, such as is caused by intensive stress or a large perturbation." In this definition, sustainability is a property of a system operating over time, a framework this chapter also advocates. The crux of the conceptual problem involves specification of the boundaries of the system and the time period, in particular as a framework for evaluating crop technology.

The problem of boundary specification arises from choosing the system level at which sustainability becomes a relevant characteristic. Much of the confusion in the discussion of sustainability reflects a mixing of system levels, namely, the lack of recognition that a plant photosynthetic system is embedded in a plant system, which is embedded in a cropping system, which is part of a farming system, which lies within the international market system. Alternatively, one could mention cellular, plant, field, continental, global, and even solar systems. Except for the highest system level—the international market— each of the lower systems is, except under quite special circumstances, open to influences from outside. Openness creates the very difficult problem of determining when sustainability is an inherent property of the defined system. Is it dependent on endogenous system relationships, as for Conway, or is sustainability so dependent on external forces that the system level should be upgraded in order to define it adequately?

Given that sustainability is a characteristic of a system's productive performance over time, it follows that the effect of a crop technology on the sustainability of the system is measured through its effect on output—the "ability to meet needs." Technology modifies the sustainability of the system in which it is being applied; it is not an inherent characteristic of a variety, a cultural practice, or a particular input. A technology's effect on a system's sustainability is thus contingent on the specification of the system and is measured by the system's output performance over time. The output measure depends on the system level. At the crop variety level it is yield per plant or per hectare; at the crop level or the cropping system level it is total factor productivity; at the farming system level it is income; and at the market level, commodity supply. We define sustainability as the capacity of a system to maintain output at a level approximately equal to or greater than its historical average, with the approximation determined by its historical level of variability. Hence, a sustainable system is one with a nonnegative trend in measured output; a technology adds to system sustainability if it increases the slope of the trend line. This definition differentiates sustainability from stability, which is the variation in the output measure around the trend line.

Measuring the slope of the trend line of a system's output involves specifying the system, the measure of output, and the time period of concern, and observing the measure over the specified time period. It will be obvious that empirical tests of sustainability will be costly and therefore applicable to only a few elite technologies within a few systems—raising an interesting issue of the value of the information in relation to the experimental costs. Moreover, none of the three specifications is straightforward.

Selection of the system level requires a choice (and therefore a trade-off) between the number of alternatives that can be screened and the range of exogenous variation to which those alternatives are exposed. The lower the system's level, the fewer the number of potential system interactions and the less complex the experimental design. One issue, then, is the lowest system level at which sustainability can be defined. Moreover, evaluation criteria at lower levels must be compatible with criteria at higher systems levels; for example, increased yields or more efficient input use must translate into higher profitability as well as output that better satisfies' market demands. At the plant level, sustainability is measured as a yield trend line, and the problem is in defining the conditions under which sustainable yields at the plant level lead to sustainable productivity at the cropping system level, which leads to sustainable incomes at the farming system level, which lead to sustainable commodity supplies, and so forth. But above the farming system level, so many factors outside the system affect its sustainability that it is virtually impossible to determine the source of such impacts. The implications of the above discussion can be summarized in a first proposition: *Sustainability is a relevant criterion for evaluating agricultural technologies only when a system using a technology has been well specified; and therefore in most cases the criterion cannot be empirically applied above the farming system level.*

The second specification problem, interrelated with the choice of system level, is the definition of output. Where crop output is evaluated under a fixed set of inputs over time, crop yield per hectare is the appropriate output measure. On the other hand, agronomic yield trials conducted over time often involve changes in specific inputs like fertilizer materials or insecticide compounds when these are not the "test" variables, often in order to protect the crop against unexpected or changing conditions. Output measures of cropping systems, which by definition include several crops, must use some means of adding together different crops, perhaps fodder and grain or fuelwood and fruit. This leads to the second proposition: *The appropriate measure of output by which to determine sustainability at the crop, cropping system, or farming system level is total factor productivity, defined as the total value of all output produced by the system during one cycle divided by the total value of all inputs used by the system during one cycle of the system; a sustainable system has a nonnegative trend in total factor productivity over the period of concern.* The value of inputs and output must be computed at a set of standardized prices which should reflect their long-term economic value.

The third specification problem is defining the time period of concern—a sufficient length of time over which the sustainability of a system can be determined. Cost considerations imply a need to delimit a sufficiently short time period to provide a projection of system output into the future with a sufficiently low probability of error. The time period of concern is clearly greater than one crop season, in nearly every case greater than three to five years, and perhaps greater than ten to twenty years. However, we believe that most decision-makers, even those concerned about the distant future, would choose a time period of less than twenty years.

Using the analogous, but far simpler, case of yield stability, it is apparent that such time-dependent parameters are rarely measured without prior information both to improve the value of the experimental information and to reduce the cost. For example, yield stability is tested in relation to available information on rainfall variability over time. This leads to the third proposition: *Sustainability of a system cannot be feasibly measured without a prior determination of the factors likely to make that system unsustainable.* An agricultural research program needs a prior determination of how a technology could lead to an unsustainable system. This leads to a very difficult chicken-egg issue, which in the physical sciences has motivated development of a theoretical structure leading to hypothesis testing and model simulation. Agricultural experimentation is only hesitantly moving in this direction.

Sustainability, Technology Design, and Ecology

Ensuring that technologies arising from crop research do not lead to unsustainable systems will require more than the mere development of testing procedures. Empirical tests are too complex and costly to implement in more than a handful of cases, so the capacity to measure sustainability at the testing and evaluation stage of the research process will be limited. Therefore it is necessary to incorporate sustainability as an objective in research planning and technology design.

One viewpoint in the sustainability debate holds that high-industrial-input agricultural systems are inherently unsustainable. Proponents of that view have shifted the focus of debate away from production or income distribution to environmental degradation and input use. Agroecology, as a scientific discipline, has led the critique of developed-world agricultural innovation over the postwar period and has been at the forefront in the formulation of alternatives to this "high-input industrial" model (Douglas, 1984). Moreover, agroecology supports the view that the high-input model is inappropriate for agricultural development in the tropics, giving as a principal reason that tropical farming systems based on such a model are "unsustainable." Sustainability, according to this view, "requires new directions for agricultural development, directions based on the principles and practical knowledge of ecology" (Dover and Talbot, 1987:7). This dichotomy has tended to politicize the term *sustainability* and associate it with a particular research agenda based on ecology.

In contrast, our concerns about the sustainability of tropical agricultural systems are directly related to how these systems will meet the increasing demand for food over the next five, ten, and twenty years, especially in countries whose population is still growing rapidly. Agricultural research having sustainability as a criterion will aim to understand how inputs and outputs are modified in systems in order to increase agricultural production and whether those shifts result in systems that give sustainable growth over time. In a dynamic environment, new agricultural technologies can facilitate such shifts, and in an otherwise static environment they may precipitate system changes. Agricultural research programs that use sustainability as a design criterion will give higher priority to research on farming systems experiencing difficulty in making a sustained adjustment to higher productivity levels. This is fundamentally different from evaluating how new technologies may affect the sustainability of the system. Design of technologies would thus increase the priority given to elements in the system that are degrading as a result of more intensive exploitation, almost invariably either some aspect of the soil or a pathosystem (Robinson, 1976).

Designing sustainable farming systems therefore requires an understanding of the process by which farmers adjust to a changing external environment, whether that is induced by climate, market expansion, or a growing population density. Farmers' ability to develop more intensive systems has been widely documented. The classic example is the farmer-initiated irrigation systems of Asia; others include farmer development of varieties; farmer initiation of varietal exchange; crop substitution, such as the substitution of cassava for yams in West Africa in response to declining soil fertility; and the development of mulching, fallow, and burn systems to maintain soil fertility (Binswanger and Pingali, 1987). Paul Richards (1986) adopted the more extreme view that in order for agricultural research to enhance the development of sustainable agricultural systems in West Africa, it should principally strengthen this existing capacity of farmers to develop their own technological solutions to changing needs.

However, we believe that population growth can be so rapid, climate change so abrupt, or market penetration so quick as to stymie the ability of farmers to adapt. Research to improve farmers' ability to sustain production will be necessary, judging from many case studies that have documented the inability of traditional farmers to intensify production at a sufficiently rapid rate to meet increasing demands on the farming system. Examples extend from Geertz's classic case of agricultural involution on Java to soil erosion in the East African highlands to desertification in the Sahel. A fourth proposition thus follows: *Whether sustainability should be a criterion of research programs depends on their target area; unsustainability is often locally or regionally defined and depends on such factors as the rate of increase in exogenous demand on the system, agroclimatic environment, and the relative intensity (generally in land use) of existing sys-*

tems. Targeting thus becomes a key issue for agricultural research programs where sustainability is a principal objective.

Ecology, with its focus on ecosystems, has extended these concerns to higher system levels. But at higher levels than the farming system, sustainability is fraught with a series of definitional problems. The boundary of the system becomes hazy, and even if it is defined, the system will likely be expected to meet several simultaneous criteria, not just sustainability. The concerns at levels above the farming system usually encompass environmental degradation and may arise from overexploitation of a common property resource and production externalities.

The "tragedy of the commons" arises in a situation of increasing population density (either human or animal) where decisions that individually have imperceptible ecological impact collectively can have substantial impact. Traditional regulatory mechanisms (collective decision-making) may be effective protection against the tragedy for peasant societies, but the protection may break down in the face of rapid population growth or expanded market opportunities. Such common resources may be grazing areas, water resources, firewood, fishing areas, or forest areas. The definitional problem—and thus how the research problem is conceptualized—can be illustrated by the issue of forest resources management. First, there is debate over the system boundary: is it ecologically defined (an agroecological region), economically defined (the region serving a timber market), or defined by a land-use system (a managed forest)? Second, what is the output of the system? The alternatives span the range from conservation (nonuse) to various logging alternatives (which can be organized for sustainable output but usually entail a loss of ecological diversity). Both problems suggest a fifth proposition: *Sustainability of common resource systems necessarily incorporates value judgments on multiple criteria over how the community wishes to utilize the resource; moreover, sustainability of the system will depend more on social institutions controlling access and use than on production technologies.*

Exacerbation of production externalities—an economic term which signifies a cost or return arising from a production activity that is borne not by the producer but by other members of society—can also arise from intensification. Soil erosion from upland farming systems that affects irrigation downstream and the leakage of agrochemicals into the environment are classic examples.

Externalities are particularly complex problems arising from three mutually reinforcing aspects of some phenomena: physical, economic, and temporal. As Kneese and Schultze (1975) emphasized, externalities are due in part to the physical conservation of matter. Pesticides applied to a field must go somewhere, usually into the groundwater; even pesticides that break down yield constituent components that go into the groundwater. If enforceable property rights exist for the groundwater, the owner can charge for the disposal of the pesticide and thereby offset its costs (i.e., internalize the costs to the producer). Externalities may also be caused by lack of information. Ignorance of pollution or of the consequences of pesticide pollution is sufficient for that pollution to exist even if a market for groundwater exists. Agricultural production processes require time, and the external effects of actions at one point are not known until some later time. If private benefits are realized early and social costs late in time, any positive social discount rate reduces the present value of costs, thereby contributing to the externality. The common property nature of resources also can generate externalities: If one individual owns both the groundwater and the land on which a pesticide is used, the owner's decisions will reflect the effect of the pesticide on groundwater quality.

Resolution of externality problems may involve intervention by the state. Moreover, their effect on a system's productive capacity usually affects several aspects—for example, value of agricultural production and environmental quality—leading to difficulties in how to evaluate alternative system states, especially where there are trade-offs.[2]

An examination of issues related to agrochemicals—inorganic fertilizers, herbicides, and pesticides—illustrates many dimensions of this conceptual problem, because such input use is at the heart of the debate over alternative agricultural development strategies.[3] The starting point is to recognize that agriculture is an extractive activity

[2] We would argue that the weights to evaluate such trade-offs should reflect long-term values to society. In a real sense, agreement on the appropriate weights is where economists and ecologists most differ, with the former arguing for weights reflecting prices that would prevail in a well-functioning market where the value of every input and output is internalized to the decision-maker, and the latter arguing for a set of weights reflecting long-term preservation of an ecologically defined ideal state. This ideal is most often reflected in preservation of the current status quo of an ecosystem. However, neither economists nor ecologists can make a convincing case for the superiority of their set of weights over the other's.

[3] Increased input utilization has sometimes been seen as synonymous with agricultural development. Johnston and Kilby (1975) provided probably the fullest account of this view, in which agrochemicals provide a basis for increasing marketed surpluses in the agricultural economy—indeed increased input use is usually first found in cash or export crops, even in peasant agriculture—and in turn provide important backward linkages for industrial development in the economy.

(Loomis, 1984) that relies on managing or controlling the crop environment, including maintenance of the soil resource. Increasing productivity, especially crop yields, increases the rate of extraction of soil nutrients at the same time that it involves increased control over the crop environment. The rate of breeding progress is in turn usually linked to these improvements in managing the crop environment. Fertilizers increase plant nutrient availability and along with other agrochemicals provide better control of the crop environment. The enhanced environment provides increased output on intensively farmed land and reduces the pressure to expand to new land.

By the late 1980s agrochemicals were an established feature of the agricultural systems of all developing countries, albeit used at widely differing intensities. What, then, are the disadvantages of a reliance on agrochemicals, and what are the alternatives? By any standard, the postwar record of U.S. crop yields and the post–Green Revolution rice yields in Asia have exhibited sustained and rapid growth. This growth has been closely linked to increases in the use of agrochemicals. Concerns about the sustainability of production levels in these farming systems are based on ecological theory about what could happen to output rather than on what has happened, and on external costs entailed in the use of pesticides and nitrates. While much of the evidence is marshaled around the not insignificant externality problem, proposed solutions often do not address the externalities but rather resort to a theory of sustainable agroecosystems. The lack of empirically demonstrated nonsustainability in part results from the time span necessary to measure sustainability. Reganold et al. (1987) showed differences in erosion losses between an organic and a conventional farming system after a thirty-seven-year period, but they estimated that actual productivity differences would not appear for another fifty years. Theory is thus a necessary part of the sustainability debate, but as Loomis's (1984) critique demonstrates, there needs to be a complementary capacity in place to test that theory. The debate is currently polarized because there are not enough data to reject or modify the two competing theories, which often verge on ideologies.

The principal alternatives to agrochemical use are improved soil quality and enhancement in the rates and efficiency of existing biological processes that control nutrient cycling and pest populations. This leads to research on topics such as enhanced biological nitrogen fixation, more efficient mycorrhiza strains, ecosystem mimicry in cropping patterns, crop rotations and multiple cropping, and biological control of insects, diseases, and weeds—topics that, even in the 1970s and 1980s, formed a considerable portion of the agenda at most international agricultural research centers.

There are several large hurdles for technologies based on such a research agenda. First, such technologies are environmentally sensitive and require in situ adjustment. The demands on research capacity will necessarily be larger than has traditionally been the case.

Second, the demands made on farmer knowledge and management will likely be greater than with high-external-input technologies. Inputs embody research knowledge and are relatively undemanding of additional farmer knowledge, while highly productive farming systems using few external inputs are labor and management intensive— at least the few that have been demonstrated. To a significant extent, this is why agrochemicals are increasingly utilized in tropical countries. On the other hand, once an improved agroecosystem is in place, it may require less management due to self-regulatory mechanisms. The difficulty is in designing and precipitating the system change.

Third, farmers will decide between using biological technologies or agrochemicals based on their knowledge, the local market economy, and government policies. Even when sustainability is an objective, farmers' choice of a technology or a change in management practices will largely be determined by its contribution to their welfare as reflected in current or near-term future profit or costs. Technologies have to increase the profit or the farmer's perceived welfare before they will be adopted and thereby have an opportunity to contribute to system sustainability. Translating the ecological research agenda into adoptable technologies that compete effectively with agrochemical alternatives will not be easy. These arguments are mustered to support a sixth and slightly contentious proposition: *Dividing research solutions to the sustainability problem into two distinct and competing strategies is counterproductive; to be successful the biological research agenda will have to complement the continued use of inputs in the intensification of farming systems in the tropics.*

Clearly, intensifying agriculture by high input use is not always the appropriate solution, as illustrated by research on intensification of shifting cultivation by IITA's farming systems program in West Africa and the INIPA–North Carolina State University program in Yurimaguas, Peru. In those projects, the objective of sustainable increase in per-hectare yields in continuous cropping systems have been achieved through high labor and external input use. Profitability has not been generally achieved, however. Interestingly, Binswanger (1986) argued that such efforts may be misplaced without a recognition of the effective demand for technology. In particular, he argued that concentrating research effort on yield increasing technologies makes little sense in the more land abundant environments. In his view, either quality-enhancing or stress-avoiding technologies would be more likely to be adopted. In recognition of these realities, the Yurimaguas group has retreated from its high-input, continuous-cultivation system (Sanchez et al., 1982) to a lower input system that uses a kudzu rotation for fertility maintenance and weed control (Sanchez and Benites, 1987). This system, nevertheless, is still labor and input intensive compared with the shifting-cultivation system, and its prospects for adoption remain undetermined. In such land-abundant conditions, biological technologies would be highly competitive with input-based solutions if they focused on enhanced stability of the system without increases in labor requirement. Ruthenberg (1980) and Binswanger and Pingali (1987) documented the pattern of intensification in sub-Saharan Africa. In most cases the evolution involves enhanced management of biological processes, which requires increased labor. To make the point again, the successful design of technologies to enhance system sustainability will begin with clearly characterized resource, farming, and marketing systems. The real utility of agroecology will come from its capacity to understand and predict system evolution, a capacity that will require some marriage of agroecology and economics.

Whether there can be a merging of the ecological perspective with the economic perspective depends, among other things, on ecologists' philosophical view of the role that markets and social institutions play in system sustainability, from the farming systems level up. Economists believe that today's world is committed to a path in which population growth, resource constraints, and human needs and desires lead to market development, which in turn leads necessarily to a division of labor, output specialization, and increased interdependence between economic agents. This implies a loss of crop diversity at the farming systems level (although diversity may be maintained at a national or a regional level) and, with the advent of input markets, a more difficult environment in which to promote the low-external-input biological alternative. Market development thus appears to undermine a significant part of the biological research agenda. But viewed slightly differently, increasing market dependence enhances the sustainability of farming systems and regional or national food systems by reducing pressure on some agroecologies while increasing output from more robust ones. Recent history, in much of Asia, for example, suggests an imperative toward increased social organization and market development in order to accommodate increasing population pressure on a limited land resource. A division of labor and trade leads to an enhanced productivity of the overall system, even without any necessary change in underlying production technology.

Moreover, trade and institutional development enhance the sustainability of food systems in important ways. As has been noted, institutional and social innovations are key to solving the problem of overuse of common resources. However, famine is probably the ultimate indicator of the unsustainability of a food system. Famines are more common in rural areas than in urban areas, and in rural areas they are more likely in regions not integrated into market systems—certainly this is the case in sub-Saharan Africa. Trade and stock management are buffering mechanisms for marginal agroclimatic regions and in a sense preserve farming systems in regions where they could not exist independently. This observation does not contradict A. K. Sen's (1981) work on famines, which suggests the fundamental role of incomes and distributional mechanisms, especially in Asia. But sustainability of food systems also depends on the appropriate design of institutions that correct for "entitlement" problems. This leads to a rather interesting and perhaps unsettling seventh proposition: *Sustainability is first defined at the highest system level and then proceeds downward; and as a corollary, the sustainability of a system is not necessarily dependent on the sustainability of all its subsystems.*

Implications for International Agricultural Research

How should agricultural research, in particular the international centers, respond to the call for sustainable agricultural systems? For a start, they should recognize the value of sustainable cropping, farming, agricultural, and national economic systems. Second, they should define appropriate measures of total productivity and establish methods for measuring it for well-defined plant systems, cropping systems, and farming systems. Third, they should conduct research with the objective of understanding the likely trend of total productivity in well-defined cropping and farming systems over appropriate time and space dimensions. Fourth, they should identify the externalities associated with such well-defined systems; and, finally, they should begin to develop methods to measure such externalities.

The first step, recognition of the issue, probably has been accomplished in all but the most recalcitrant centers, although many individuals at the centers are more cautious about sustainability than the international trendsetters among the donor community. After all, the latter group can simply issue the call for sustainability research and go back to their business of seeing how responsive researchers are. The researchers have to determine how to measure the concept so that the work carried out under the banner adds to knowledge that will enhance food production, small-scale farmers' incomes, consumers' welfare, women's status in development, national research program capacity, stability, *and* sustainability of agriculture.

Moreover, it is likely that some agroecologies are inherently suitable for intense use, while others can be used sustainably only at low levels of intensity. Successful agricultural research will stratify its target areas accordingly and devise ways to raise intensity levels on the former, thereby permitting the latter to sustain low levels of output. More concretely, agronomic researchers conducting field experiments on production systems and economists who analyze systems research must move from a fixation on yield as the measure of agricultural system success to a measure of total productivity. This implies, as discussed earlier, a definition of total productivity that includes, as much as possible, all costs and benefits, not just those accruing in the immediate time period and to the immediate decision-maker. Like all ambitious goals, this objective will not be easy to achieve, but one can approach it incrementally. Comparable total productivity measures require measurement of all inputs and outputs of experimental systems, agreement on weights, and computation of comparable total productivity measures for alternative treatments of the system. This is clearly within the reach of the international agricultural research centers and should be a priority of their agronomic and economic researchers. Identification and costing of externalities is more difficult; it requires specific methodologies that are still being developed (Antle and Capalbo, 1988). It will be impossible, however, unless the first step is initiated.

The third step that international research centers could take is to institute a set of long-term cropping or farming system experiments in a limited number of agroecologies typical of their mandate responsibilities. These would have to be carefully designed to provide useful information in the short run as well as over the long term. They would have to contain a sufficient range of treatments of a system to provide a set of comparisons that would be useful as economic and weather conditions changed. Plots would have to be large enough to reflect what might happen under local farming conditions and permit the possibility of future modification of treatments. One might build in flexibility by designating certain treatments the "best commercial practice under current prices," or "most promising new cultivar," or "integrated pest management." A careful advance assessment of the likely sources of perturbation will be necessary if the experiments are to be useful in making judgments about sustainability and stability. And one must recognize that the experiments, however carefully designed, will be inadequate for many purposes. They may not identify sustainable systems, but at the end of ten or twenty years, a great deal more information and insight will be available to examine the issues of sustainability *and* productivity *and* stability *and* input use, and much else that is now unknown.

The fourth step links directly to the third—scientists should examine the experimental systems for all possible externalities and identify them. The fifth step, developing methods to measure and value such externalities, then follows directly. These efforts will require new skills and techniques and hence may take most agricultural research organizations some time to accomplish, but they should be started now.

While the above may be viewed as the necessary first steps to addressing sustainability as a research objective, they do not grapple with the issue of how international agricultural research adapts or organizes programs so that sustainability is addressed. The alternative approaches are three: (1) a recasting of existing commodity research programs, that is, a continued focus on plant-breeding research with the incorporation of the five additional steps outlined above; (2) organizing research around resource management, for example, around soils, irrigation, or forestry; and (3) organizing research around "solutions," for example, agroforestry, tropical soil biology, and insect physiology. There is an emerging structure in the form of international research networks and centers organized around these latter two disparate approaches. Moreover, CGIAR centers are adding or adapting research programs along similar lines. These are in most cases independent efforts, which leads to the natural question of whether there is an emerging order in this "rush" to sustainability.

This emerging order will have to resolve two fundamental difficulties that characterize sustainability research; namely, how to organize research on problems whose solutions are very location-specific, and how to organize biological research where the focus is on the whole agricultural system rather than individual components. At issue is whether some integration of the three above approaches can overcome or accommodate these difficulties.

Researching whole agroecosystems is not practical. First, it is not clear how to select and evaluate alternative states of an agroecosystem so that the biological performance of the whole system is examined under alternative treatments that can be compared and evaluated. Second, the number of agroecosystems in the tropics is essentially infinite, precisely because these managed biological systems are so finely tuned to the great variation in soils, climate, pest complexes, resource availability, and output markets. System definition to focus the research under such circumstances becomes impossible. Alternatively, sustainability research can be structured around components within an integrating framework, even when the broad objective is commodity improvement. Agroecology provides a theoretical framework for the selection of these components, and a number of research groups, and some commodity research programs, are working on biological nitrogen fixation, integrated pest management, biological pest control, agroforestry, tropical soil biology, and multiple cropping.

Alternatively, a significant amount of applied biological research has been organized around management of natural resources—for example, in forestry and fishery management. Such research focuses on the management of a single natural resource system where the biological yield of the system largely coincides with the economic yield. However, when agriculture, soils, water, and forests are managed as parts of farming systems, there is often a disjunction between management of the sustainable productivity of a single resource and organizing output and input mix so as to optimize income. The difficulty for organizing sustainability research—if not the organizational paradox—is suggested by the example of hillside maize systems: the erosion problem cannot by solved by a singular focus on the maize system, and in turn, a singular focus on erosion control technology without consideration of the maize-cropping system is impossible. Equally, research to optimize management of Vertisols or acid soils cannot be done without considering alternative cropping systems to be grown on those soils. Research on resource management is, in reality, component research. The difference is in the definition of components and their coincidence with farmers' objectives.

Thus, the systems problem inherent in sustainability research is being addressed by research groups that focus on components defined as subsystems, resources, or commodities. Such component research faces the challenge of developing solutions of general applicability that meet the individual requirements of particular systems. This problem is usually resolved through adaptive research that integrates components into systems, taking as its research the long-run trend of total factor productivity. The basic and applied research objectives that underlie the development of sustainable systems are met by focusing on subsystems. Nevertheless, it is incumbent on researchers in these programs to be informed about research on other components and how the subsystems might interact. That is, they should be doing component research with a sustainability perspective.

Putting improved systems together relies on two complementary approaches: system simulation models (Dent and Thornton, 1988) and adaptive research organized around agroecosystems (Conway, 1985). Much like farming systems research, the application of agroecosystems as a research methodology has greatest utility at the adaptive research level. Whether it also can be a bridge to problem identification for applied research remains to be seen—farming systems research has in practice not performed as well as its designers had hoped as a vehicle for information feedback to basic and applied research programs (Lynam, 1986).

The CGIAR system is now assessing how it can best implement sustainability as a research objective. Two distinct but not necessarily contradictory approaches are emerging. TAC, in its policy paper (CGIAR, 1988), delegated responsibility to each of the centers in the sense that each has to incorporate the sustainability objective into its work. How this will be done is left to the individual centers' strategic decision-making processes. However, the TAC paper recognizes in a vague way that such an incorporation will require some reorganization and a reallocation of staff to "resource management." This reorganization is a tacit admission that the sustainability problem must be broken down into a set of researchable component areas that complement the commodity research programs. What those areas are and which centers will adjust their programs to include them are issues left in abeyance.

Another proposal (Colmey and Schuh, 1988) suggests that CGIAR could meet the sustainability objective by incorporating additional centers that address resource management. Such an expansion would ensure that CGIAR addresses the sustainability objective only to the extent that centralized resource allocation also results in better inter-center collaboration, a more rational division of labor, and more effective priority and program definition. Such coordinating mechanisms presently exist only weakly within the CGIAR system. A concerted move to allocate the various dimensions of sustainability research among (existing and additional) centers would require a radically stronger coordination mechanism. Most observers agree that one of the strengths of the CGIAR system is its relatively unstructured nature. Must the organization of CGIAR become more complex to deal with the sustainability problem, leading to the danger that what it gains in coordination of activities it will lose in bureaucratic rigidity and information management costs?

Summary and Conclusions

We agree with those who are concerned that agricultural production systems should be sustainable, and further, we believe that technology can be designed to contribute toward increasing the sustainability of systems. We have developed the following propositions which provide a framework in which international agricultural research centers can empirically address sustainability:

- Sustainability is a relevant criterion for evaluating agricultural technologies only when a system using a technology has been well specified; and therefore, in most cases, the criterion cannot empirically be applied above the farming system level.
- The appropriate measure of output by which to determine sustainability at the crop, cropping, or farming system level is total factor productivity, defined as the total value of all output produced by the system during one cycle divided by the total value of all inputs used by the system during one cycle of the system; a sustainable system has a nonnegative trend in total factor productivity over the period of concern.
- Sustainability of a system cannot be feasibly measured without a prior determination of the factors likely to make that system unsustainable.
- Whether sustainability should be a criterion of research programs depends on their target area; unsustainability is often locally or regionally defined and depends on such factors as the rate of increase in exogenous demand on the system, agroclimatic environment, and the relative intensity (generally in land use) of existing systems.
- Sustainability of common resource systems necessarily incorporates value judgments on multiple criteria over how the community wishes to utilize the resource; moreover, sustainability of the system will depend more on social institutions controlling access and use than on production technologies.
- Dividing research solutions to the sustainability problem into two distinct and competing strategies is counterproductive; to be successful the biological research agenda will have to complement the continued use of inputs in the intensification of farming systems in the tropics.

- Sustainability is first defined at the highest system level and then proceeds downward; and as a corollary, the sustainability of a system is not necessarily dependent on the sustainability of all its subsystems.

The minimum steps that international agricultural research centers should take to address agricultural system sustainability are (1) to recognize the need for sustainable agricultural systems, (2) to define appropriate ways to measure sustainability, (3) to empirically examine the sustainability of some well-defined cropping or farming systems, (4) to define the externalities that exist in such systems, and (5) to develop methods to measure those externalities. Even while recognizing that sustainable agriculture requires more than sustainable farming systems, we believe that these steps will begin to generate knowledge that will lead to that larger goal.

The supposed advantage of organizing research with a sustainability objective around resource management or "solutions" instead of around commodities is not at all evident. Organizing multidisciplinary agricultural research along commodity lines has been successful in producing new technologies principally because of the correspondence between researchers' and farmers' evaluation criteria. How resource management and solution centers will organize their research and interact with commodity research programs to produce sustainable technologies will be determined as all these centers implement research programs that measure the sustainability of various agricultural systems.

Bibliography

Antle, J. M., and S. M. Capalbo. "Pollution Externalities and the Economic Evaluation of Agricultural Technologies." Paper prepared for the Agricultural Sciences Division of the Rockefeller Foundation, Washington, D.C., 1988.

Binswanger, H. "Evaluating Research System Performance and Targetting Research in Land-Abundant Areas of Sub-Saharan Africa." *World Development*, 14:469–75, 1986.

Binswanger, H., and P. Pingali. "The Evolution of Farming Systems and Agricultural Technology in Sub-Saharan Africa." In V. W. Ruttan and C. E. Pray, eds., *Policy for Agricultural Research*. Boulder, Colo.: Westview Press, 1987.

CGIAR (Consultative Group on International Agricultural Research). *Sustainable Agricultural Production: Implications for International Research*. Rome: TAC Secretariat, Food and Agriculture Organization of the United Nations, 1988.

Colmey, J., and G. E. Schuh. "Natural Resources and Sustainability in the Developing Countries: Meeting the Challenge with International Research." Mimeo, April 17, 1988.

Conway, G. R. "Agroecosystem Analysis." *Agricultural Administration,* 20:31–55, 1985.

Dent, J. B., and P. K. Thornton. "The Role of Biological Simulation Models in Farming Systems Research." *Agricultural Administration and Extension,* 29:111–22, 1988.

Douglas, G. K. *Agricultural Sustainability in a Changing World Order.* Boulder, Colo.: Westview Press, 1984.

Dover, M., and L. M. Talbot. *To Feed the Earth: Agro-ecology for Sustainable Development.* Washington, D.C.: World Resources Institute, 1987.

Holling, C. S. "Resilience and Stability of Ecological Systems." *Annual Review of Ecology and Systematics,* 4:1–23, 1973.

Johnston, B. F., and P. Kilby. *Agriculture and Structural Transformation: Economic Strategies in Late-developing Countries.* New York: Oxford University Press, 1975.

Kneese, A., and C. L. Schultze. *Pollution, Prices, and Public Policy.* Washington, D.C.: Brookings Institution, 1975.

Loomis, R. S. "Traditional Agriculture in America." *Annual Review of Ecology and Systematics,* 15:449–78, 1984.

Lynam, J. K. "On the Design of Commodity Research Programs in the International Centers." In D. Groenfeldt and J. L. Moock, eds., *Social Science Perspectives on Managing Agricultural Technology.* Colombo: International Irrigation Management Institute, 1989.

Pinstrup-Andersen, P., and D. Franklin. "A Systems Approach to Agricultural Research Resource Allocation in Developing Countries." In T. M. Arndt, D. G. Dalrymple, and V. W. Ruttan, eds., *Resource Allocation and Productivity in National and International Agricultural Research.* Minneapolis: University of Minnesota Press, 1977.

Reganold, J. P., L. F. Elliot, and Y. L. Unger. "Long-Term Effects of Organic and Conventional Farming on Soil Erosion." *Nature* (London), 330:370–72, 1987.

Richards, P. *Indigenous Agricultural Revolution.* London: Hutchinson, 1985.

Robinson, R. A. *Plant Pathosystems.* Berlin: Springer-Verlag, 1976.

Ruthenberg, H. *Farming Systems in the Tropics.* London: Oxford University Press, 1980.

Sanchez, P. A., D. E. Bandy, J. H. Villachica, and J. J. Nicholaides. "Amazon Basin Soils: Management for Continuous Crop Production." *Science,* 216:821–27, 1982.

Sanchez, P. A., and J. R. Benites. "Low-Input Cropping for Acid Soils of the Humid Tropics." *Science,* 238:1521–27, 1987.

Sen, A. K. *Poverty and Famines.* Oxford: Clarendon, 1981.

12

Economic Analysis of Soil Erosion Effects in Alley-Cropping, No-Till, and Bush Fallow Systems in Southwestern Nigeria

S. K. Ehui, B. T. Kang, and D. S. C. Spencer

Introduction

Shifting cultivation is typical of traditional agricultural systems in tropical Africa. Farmers fell and burn the woody vegetation, cultivate the cleared land (typically for one to three years), and then abandon the site (for four to twenty years) to forest and bush cover (Kang et al., 1985; Sanchez, 1976). Until recently, enough arable land was available to use it in this fashion. Today, however, with increasing populations, fallow periods have decreased, leading to more intensive cultivation of marginal land as well as to clearing new forest lands for agriculture (El-Ashry and Ram, 1987; Lal et al., 1986).

A major consequence has been increased soil erosion from cropland, declining soil fertility, and lower crop yields (Kang et al., 1985), primarily because of the loss of organic matter and nutrients contained in the eroded sediments. Soil productivity declines further be-

This chapter is a modified version of our article "Economic Analysis of Soil Erosion Effects in Alley Cropping, No-Till and Bush Fallow Systems in South Western Nigeria," *Agricultural Systems,* 34:(4), 1990. We acknowledge the helpful assistance of O. I. Okunola, Yemissi Oni, Tunde Oyewole, and S. Olayinka during the collection of the data. Ms. Okunola served as a valuable assistant and interpreter during the interviews. Wade Brorsen, Tim Baker, Tom Hertel, and John Lynam provided valuable comments on an earlier draft of this paper.

225

cause soil lost through erosion reduces both the thickness of the root zone and the soil's moisture-holding capacity (Lal, 1985; Lal and Kang, 1982).

Where soil loss rates are lower than the rates of soil formation, soil productivity can be maintained. However, most upland soils in humid and subhumid tropical Africa cannot sustain continuous production under current practices because they are unstable, sandy, highly weathered, low in organic matter content, and susceptible to erosion and compaction. They are washed away by frequent and intense rainfall and are easily degraded under continuous or short-fallow cultivation (Matlon and Spencer, 1984; Lal, 1985). Many tropical nations thus face a dilemma: how to feed an increasing population without irreparably damaging the natural resource base on which agricultural production depends (Ehui and Hertel, 1989; Greenland and Lal, 1977; Lal, 1986).

In its efforts to alleviate the food shortage problem in sub-Saharan Africa, the International Institute of Tropical Agriculture (IITA) has concentrated its research over the past two decades on developing sustainable soil management technologies that enhance food production and preserve the natural resource base. The most promising land-use systems are alley cropping and no-till farming (Verinumbe et al., 1984). Alley cropping is an agroforestry system in which food crops are grown in alleys formed by rows of trees or shrubs, which are periodically pruned during the cropping season to prevent shading and reduce competition with the food crops (Kang et al., 1985). In no-till farming, disturbance of the soil is kept to a minimum and crops are seeded through the residue of a previous crop or through sod without plowing. Herbicides are used to control weeds. Both systems control soil erosion in addition to maintaining soil fertility (Kang and Ghuman, 1989).

Although some analyses evaluate the economic viability of improved land-use systems in the humid and subhumid tropical regions of Africa (Hoekstra, 1982; Ngambeki, 1985; Raintree and Turray, 1980; Sumberg et al., 1987; Verinumbe et al., 1984), virtually none of them account for the erosion process and its resultant long-run impact on costs and returns. Failure to incorporate the effect of erosion's on-site costs on the farmer's land-use decision may give misleading results about the effectiveness of policy measures on agricultural production systems.

On the basis of a simulation model, this chapter uses a capital-budgeting approach to determine how these land management technologies compare with each other and with traditional bush fallow systems, taking into account the short- and long-run impact of soil erosion on agricultural productivity and profitability under alternative population density scenarios in southwestern Nigeria. This study differs from previous research in that the productivity effects of soil erosion are assessed.

Procedure and Data

Five fallow management systems are evaluated in the study: alley-cropping systems with leucaena *(Leucaena leucocephala)* hedgerows planted at 2-m and 4-m intervals, continuous no-till, and two traditional fallow management systems with 25 and 50 percent farming intensities. The 25 percent land-use intensity system is represented by a three-year cultivation and a twelve-year fallow rotation system; and the 50 percent farming intensity system by a three-year cropping in a six-year-cycle shifting-cultivation system. Maize was chosen because it is commonly grown in southwestern Nigeria; and although numerous variations in land-use systems are possible, most agronomic trials have used maize as the food crop (Kang and Ghuman, 1989; Lal, 1986).

Southwestern Nigeria (which lies roughly between longitudes 3° and 6° east and latitudes 6° and 8° north) provides a good example of areas facing erosion problems. This region is characterized by a bimodal rainfall distribution, with annual rainfall ranging between 1100 and 1500 mm. The most common soil categories found there are alfisols. Although they vary considerably, these soils have typically shallow effective rooting depths because of both argillic and gravel subsoils. They suffer a rapid decline in infiltration after cultivation and have a poor structural stability, which makes them vulnerable to erosion (Aneke, 1986; Lal, 1976).

The Capital-Budgeting Model

A multiperiod partial budget approach allows for variations over time in yield due to productivity loss. A long-run approach is necessary to capture the interaction between yield-depressing effects of soil erosion and the yield-increasing effects of improvement in fallow management systems. Assessing the long-run productivity loss from erosion may result in an increased economic incentive to adopt conservation practices even though they may compare unfavorably with traditional practices in the short run (Crosson and Stout, 1983; Walker and Young, 1986). A long-run approach also makes it possible to examine the influence of different discount rates and planning horizons on the economic attractiveness of various erosion control practices.

Virtually all economists agree that farmers' erosion control practices are based on their estimates of the net present value of their investment (Ervin and Ervin, 1982; Seitz and Swanson, 1980). The argument contends that any soil erosion that lowers productivity also lowers future returns to the land. Therefore, farmers will adopt a conservation practice if the rate of return to capital invested exceeds the opportunity cost of capital. This requires at a minimum that the cost of the current practice be less than the present value of the benefits.

Based on the above argument, the proposed model estimates the net present value at the end of the current crop year of the private costs and benefits accruing over a relevant time horizon from choosing a soil-conserving system over the traditional shifting-cultivation practice. The farmer is assumed to be using an erosive system and contemplating switching to a soil conservation practice (e.g., alley cropping). The present value of incremental net returns (PVINR) from investing in that practice can be written as:

$$\text{PVINR} = \sum_{t=0}^{T-1} (\pi_t - \pi_t')\beta^t + \rho\Delta SV_T\beta^{T-1} \tag{1}$$

where π_t and π_t' are the net returns per hectare in year t for the "with" and "without" technology scenarios, respectively; β denotes the discount factor and is equal to $(1 + r)^{-1}$, where r is the discount rate; ρ is the expected percentage of change in salvage value (i.e., terminal net worth) to be received by the farmer at the end of the planning horizon. If $\rho < 1$, the farmer will not receive the full benefits of the erosion control investment. This is expected because of the imperfect land market that prevails (Ega, 1985). When land is sold, presumably farmland market participants underestimate the effect of erosion on future productivity. The case that $\rho = 1$ is a naive approach indicating a perfectly functioning land and capital market with buyers of farmland having sufficient information about the land they purchase (Ervin and Mill, 1986).

ΔSV_T is the change in the private salvage value owing to the new technology and is calculated as:

$$\Delta SV_T = \sum_{t=T}^{\infty} (\pi_t - \pi_t') \beta^{t-T} \tag{2}$$

Assuming that π_t and π_t' stay constant after $t = T$ (the economy being thus forced into steady state), the infinite horizon model can be converted into a finite horizon equivalent by giving a weight of $\beta^T/(1-\beta)$ to the change in the private salvage value. If PVINR > 0, the farmer gains from choosing the soil-conserving system. However, if PVINR < 0, the farmer would have the incentive to continue mining the soil by using the traditional erosive practice. In this case, a subsidy at least equal to the PVINR may be necessary to override the private advantage of using the erosive practice (Walker and Young, 1986).

The central feature of equation 1 is the specification of the net return functions, π_t and π_t'. Both are defined as price times yield less the variable costs of production:

$$\pi_t = P_t Y(Z_t) - C_t \tag{3}$$

$$\pi_t' = P_t Y(Z_t') - C_t' \tag{4}$$

In equations 3 and 4, $Y(Z_t)$ and $Y(Z_t')$ are the crop yield functions for the "with" and "without" soil-conserving systems, respectively. They are assumed to depend upon cumulative soil loss rates

$$Z_t = \sum_{t=0}^{t-1} d_\tau$$

and

$$Z_t' = \sum_{t=0}^{t-1} d_\tau'$$

where d_τ and d_τ' denote the amount of topsoil lost in year t for the conservation and erosive practices, respectively. Cumulative soil loss rates are explicit arguments in yield functions because crop yield in any year is affected by the depth of the soil remaining and the available technology (Lal, 1981; Walker 1982). On the basis of past agronomic evidence, we assume that yields will decline as cumulative erosion increases, owing to nutrient losses and decreasing levels of organic matter through erosion (Lal, 1986; Young, 1986). Because productivity loss increases with cumulative soil loss, the economic incentive for using an erosive practice decreases as erosion proceeds. P_t is the price of the crop in year t; C_t and C_t' denote the variable costs of crop production with and without the soil management technology in year t, respectively.

Substituting π_t and π_t' in equation 1 gives (after rearranging terms):

$$PVINR = \sum_{t=0}^{T-1} \{P_t[Y(Z_t) - Y(Z_t')] + [C_t' - C_t]\}\beta^t$$
$$+ \ \rho[\beta^T/(1-\beta)]\{P_T[Y(Z_T) - Y(Z_T')] + [C_T' - C_T]\} \qquad (5)$$

The right-hand side of equation 5 is composed of three parts. The first bracketed term, $\{P_t[Y(Z_t) - Y(Z_t')]\}\ \beta_t$, is the present value of the yield differential between the improved (soil-conserving) and erosive systems. This expression will be positive (negative) if the soil-conserving system is higher (lower) yielding. The next component, $[C_t' - C_t]\ \beta^t$, measures the net cost of adopting the conservation practice. It thus reflects any increase (saving) in operating inputs (e.g., labor, chemical or material costs) due to the new technology. The last term, $\rho[\beta^T/(1-\beta)]\ \{P_T\ [Y(Z_T) - Y(Z_T')] + [C_T' - C_T]\}$, is a terminal value that reflects the discounted value of all incremental returns to be realized beyond $T-1$.

Data Sources and Construction

In order to determine the effects of soil erosion on net returns over time and under different fallow management systems, data on the relevant variables were collected and simulated (where appropriate) for conditions prevailing in southwestern Nigeria.

Soil Loss. Rates of soil erosion under the different fallow management systems are calculated over a twenty-year planning horizon using the SCUAF (soil changes under agro-forestry) simulation model (Young et al., 1987) in combination with soil erosion rates measured under experimental conditions at the International Institute of Tropical Agriculture. Soil erosion loss is modeled using the universal soil loss equation in its simplified form as employed by FAO (1979):

$$A = R \times K \times S \times C \qquad (6)$$

Figure 12.1. Predicted cumulative soil erosion loss under five fallow management systems for Oxic paleustalf in southwestern Nigeria

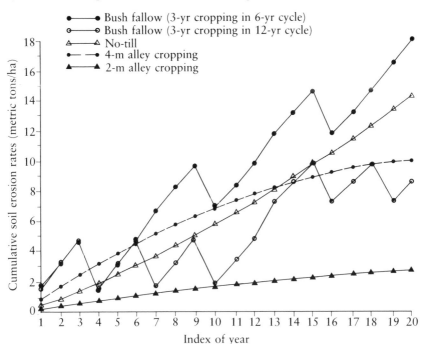

where A is the annual soil loss in kg/ha, R is the climatic (rainfall) factor, K is the soil erodibility factor, S is the slope factor, and C is the cover and management factor, which is determined by crop rotation, intercropping, and the tillage system. With these factors, erosion is calculated separately for the tree and crop components. For rotational agroforestry systems such as shifting cultivation, the values of erosion under tree and crop are used in the respective years under these components. For spatial agroforestry systems (such as alley cropping), a tree proportionality factor is defined, which represents the degree to which erosion as a whole is controlled by the tree component.

In Figure 12.1, cumulative soil loss rates are plotted over time for the five fallow management systems described above. The soil is a moderately sloped sandy Oxic paleustalf type with a mean 7 percent slope gradient. The alley-cropping system with 2-m interhedgerow spacing shows the lowest estimated cumulative soil loss rates after twenty years (2.5 metric tons/ha). It is followed by the 25 percent farming intensity bush fallow system (8.37 metric tons/ha), the alley-cropping system with 4-m interhedgerow spacing (9.9 metric tons/ha), the no-till (14 metric tons/ha), and the traditional shifting-cultivation system with a 50 percent farming intensity (17.8 metric tons/ha). Note that initially erosion rates are lower under no-till compared with the 4-m alley cropping. The latter, however, gives a relatively lower cumulative soil loss, owing to the tree proportionality factor, at the end of the planning horizon. Initial soil loss rates collected from field trials in Ibadan (southwestern Nigeria) in no-till farming and alley-cropping systems with 2-m and 4-m interhedgerow spacing are used (0.43, 0.17, and 0.82 metric tons/ha, respectively [Kang and Ghuman, 1989]). The initial soil loss rate of 1.53 metric tons/ha under the bush fallow systems is based on experimental results of Sabel-Khoshella et al. (1984).

Crop Yields. Maize yield over the twenty-year planning horizon is determined based on the predicted cumulative soil erosion rates described above. The appropriate function relating maize yield to cumulative soil loss (i.e., for a sandy Oxic paleustalf with a 7 percent slope under southwestern Nigeria conditions) is from Lal (1981):

$$Y_t = 6.70e^{-0.003Z_t} \qquad R^2 = 0.89 \qquad (7)$$

where Z_t denotes cumulative soil loss rates in year t.

With equation 7, maize yield levels under the five land-use systems are calculated for the base case, in which yields are expressed over the land occupied by the maize crop only (Figure 12.2). The 2-m alley-cropping system maintains relatively higher yields over the twenty-year planning horizon. It is followed by 4-m alley-cropping, no-till, and the twelve-year-cycle bush fallow system with 25 percent farming intensity. The six-year-cycle shifting-cultivation system with a 50 percent farming intensity gives the lowest yields owing to relatively higher cumulative soil loss rates. Initial yield levels collected from field trials in Ibadan are 3.45, 3.13, and 2.40 metric tons/ha for the 2-m and 4-m alley-cropping systems and no-till, respectively (Kang and Ghuman, 1989). For the traditional bush fallow systems, an initial yield level of 2.27 metric tons/ha is chosen, based on tilled-controlled plots. This is a reasonable approximation as maize yields under traditional conditions range between 1.0 and 2.0 metric tons/ha (Mutsaers et al., 1987; Oyo North ADP, 1989).

Because of differences in resources and management practices, yield gaps normally prevail between experiment station yields and yields

Figure 12.2. Projected maize yields as a function of cumulative soil losses for five fallow management systems over twenty years for Oxic paleustalf in southwestern Nigeria

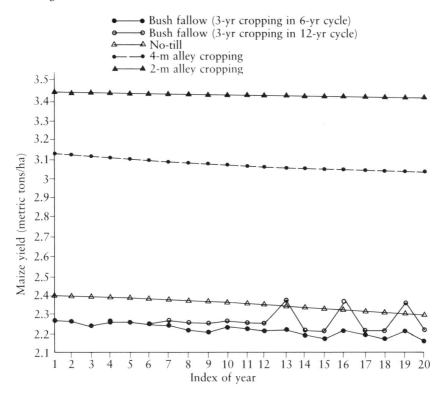

on farmers' fields. It is normal practice in economic analysis to discount the experiment station yields to account for the gaps. This method, however, is not adopted here because the predicted maize yields under the traditional bush fallow systems are based on experiment station results and are consequently higher than standard yields on farmers' fields. Contrary to some practices (e.g., Verinumbe et al., 1984), maize yields under alley-cropping systems have not been discounted, despite the fact that under these agroforestry systems land is occupied permanently by leucaena. Evidence from agronomic studies indicates that maize yield is concave in the density variable. In other words, maize yield increases but at a decreasing rate with increases in the density of the plant population. Experiment station results have shown that maize yield stagnates when population ranges between

Table 12.1. Estimated average costs (naira per hectare per year) of maize production under alternative fallow management systems, southwestern Nigeria, 1988

	Continuous cultivation			Bush fallow	
Inputs	2-m alley	4-m alley	No-till	3-yr cropping in 6-yr cycle	3-yr cropping in 12-yr cycle
Maize seeds	14.8	14.8	14.8	14.8	14.8
Herbicides	—	—	150.0	—	—
User costs[a] (equipment & tools)	73.7	73.7	139.3	73.7	73.7
Labor costs	1096.6	906.4	398.8	615.0	430.5
Imputed land clearing cost	37.5	37.5	37.5	150.0	75.0
Capital costs (imputed at 10%)	112.3	103.2	74.0	85.4	59.4
TOTAL	1344.9	1135.6	814.4	938.9	653.4

[a] The user costs of the farm tools and equipment are computed based on the capital recovery factor formula: $A = PV\{r/(1-[1+r]^{-t})\}$, where A is the annualized cost of capital item, PV is the present value of the capital item defined as the purchase price or less the present worth of its future salvage value, t is the estimated life span of the capital item, and r is the discount rate. A 10 percent discount is used for the base case, which is based on the opportunity cost of capital for Nigeria (IBRD, 1987).

40×10^3 and 60×10^3 plants per hectare (IITA, 1986). There are also arguments that penalties must be imposed on yields in the no-till fields because of increased disease, weed control, and germination problems under the "trashy seedbeds" that characterize conservation tillage (e.g., Taylor and Young, 1985). This procedure is not adopted owing to a lack of empirical evidence on the magnitudes of the penalties.

Crop Price and Production Costs. The crop price is set at 0.59 naira (N) per kilogram (in 1988, U.S. $1 = N5 on average), the average of farm-gate maize prices during the 1988 peak harvest period (June, July, and August) in Ibadan, Nigeria. Estimates of total variable costs per hectare for each fallow management system are reported in Table 12.1. They include planting expenditures (maize seeds), user costs of fixed equipment, herbicide costs (for the no-till), labor costs, and imputed capital costs obtained from a survey of 25 smallholder farmers randomly drawn from a list of 208 households in selected villages near Ibadan, Nigeria.

The average farm size is estimated at 0.94 ha, with maize and cassava as the main food crops. The average maize-planting expenditure in the sample was estimated at N14.81/ha. The most popular farm tools are machetes and hoes. On average, farmers own 2.64 machetes, with a life span of 1.16 years. The average number of hoes held is 2.44, with a life span of 3.36 years. The unit of cost of a machete is estimated at N24.2, while that of a hoe is estimated at N9.54. If herbicides are used, a sprayer is added to the stock of equipment, at a purchase price of N380 with a 10-year life span. The annual user costs of the farm tools and equipment are calculated using the capital recovery factor formula (Table 12.1). This gives N139.31/ha for the no-till system and N73.7/ha for all other production systems. The unit cost of herbicide is estimated at N150/ha. This is specific to the most popular chemical for maize, Primextra, and is based on the recommended 5 l/ha.

Annual labor requirements for maize production under the alternative fallow management systems were estimated by Ngambeki (1985). Eighty-two man-days of labor are required for the six-year-cycle traditional fallow management system with 50 percent farming intensity. In the 4-m alley system, annual labor use is increased by 53 percent owing to added labor for hedgerow pruning. In the no-till system, total labor use is reduced by 35 percent because of lower weeding requirements due to herbicide application. In the 2-m alley-cropping system, total labor required is assumed to increase by 87 percent owing to the added pruning time of the increased number of leucaena hedgerows. We estimate that in 2-m alley-cropping systems, 33 percent of the land is occupied by trees, compared with 20 percent in the 4-m alley-cropping system. In the traditional fallow systems with six years of fallow, total labor required is assumed to decrease by 30 percent as a consequence of low incidence of weed infestation. Labor costs per hectare for all land-use systems are calculated using a wage rate of N7.5/man-day, based on the prevailing minimum wage. This also reflects current wage rates in the survey area, which range between N5 and N10 per man-day.

Because bush fallow and shifting agriculture require more frequent land clearing than the continuous alley-cropping and no-till systems, it is important for comparison purposes to account for the cost of clearing land, which is calculated based on an estimated 100 man-days of required labor to clear one hectare of high bush fallow (Oyo North ADP, 1989). Assuming that all cropped lands were equally forested initially, the annual land-clearing cost in each system is computed by prorating the required total labor cost linearly over twenty years. Thus for the continuous-cultivation systems (alley cropping and no-till), the annual cost of land clearing is estimated at N37.50/ha per year because land is cleared only once. For the traditional system with 25 percent farming intensity (i.e., three-year cropping in a twelve-year-cycle bush fallow system), the annual cost of land clearing is estimated at N75/ha per year because 400 man-days of labor are required to clear 4 ha of land during the first twelve years of the planning horizon. In the remaining eight years, only 3 ha of land are cleared, which requires 300 man-days. Following the same procedure as above, the imputed annual land-clearing cost for the traditional system with 50 percent farming intensity (i.e., three-year cropping in a six-year-cycle bush fallow system) is estimated at N150/ha per year. In this case, 600 man-days of labor are needed to clear 2 ha of land during the first eighteen years of the planning horizon. For the remaining two years, only 1 ha of bush fallow is cleared, which requires 100 man-days.

Empirical Results and Discussion

Assuming that the farmer has been using the traditional shifting-cultivation system with a 50 percent farming intensity, the income effects of soil erosion under alternative fallow management systems are assessed by computing the present value of incremental net returns (PVINR) over a twenty-year planning horizon (see eq. 1). These results are reported in Figure 12.3 for the base case (scenario 1). A discount rate of 10 percent is used, based on the World Bank's estimate of the opportunity cost of capital for Nigeria (IBRD, 1987). The opportunity cost of land is assumed to be limited, reflecting the case of a low population density. Assuming the farmer does not receive anything from the change in salvage value (i.e., $\rho = 0$), the traditional bush fallow system with a 25 percent farming intensity is the most profitable, with a PVINR of N2546.8/ha, followed by the 4-m alley-cropping, the no-till, and the 2-m alley-cropping systems, with PVINRs of N1755.0, N1712.1, and N1373.3/ha, respectively.

We estimate that annual labor costs must decrease by at least 10 and 12 percent for the 4-m and 2-m alley cropping to be competitive with the twelve-year-cycle bush fallow systems, ceteris paribus. Therefore, where land is abundant and access to new forest land is "cost-less," the attractiveness of soil-conserving technologies such as alley cropping or no-till is limited compared with bush fallow systems with long fallow periods (e.g., nine years). This reinforces the argument that in land-abundant countries, strategies based on area expansion are the lowest-cost sources of growth (Pingali et al., 1987).

Figure 12.3 indicates that the present value of incremental net return from switching to any of the fallow management systems in-

Figure 12.3 Per-hectare present value (discount rate = 10 percent) of incremental net returns of four alternative fallow management systems compared with a three-year cropping in a six-year-cycle bush fallow system in southwestern Nigeria

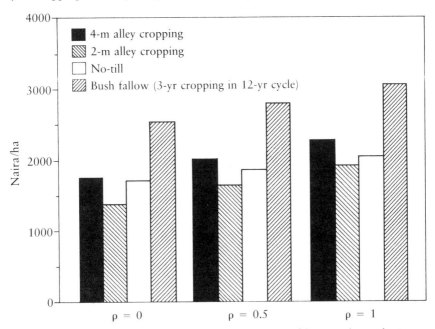

Note: No yield penalties are imposed, reflecting the case of low population density. ρ denotes the expected percentage of change in terminal net worth to be received by the farmers.

creases as the amount of the salvage value the farmer expects to receive increases. Thus, assuming, for example, that the farmer will receive 50 percent of the change in salvage value, the present value of incremental net returns per hectare increases to N2810.1, N2020.8, N1881.4, and N1652.4 for the twelve-year-cycle shifting-cultivation system, the 4-m and 2-m alley-cropping systems, and the no-till system, respectively. Note in particular that the profitability of the 2-m alley-cropping system increases faster than that of the no-till, reflecting a higher net benefit of the former compared with the latter at the end of the twenty-year planning horizon. In the interest of brevity, subsequent analysis focuses only on the case of $\rho = 0$.

Figure 12.4 reports results of a second scenario, taking into account the effect of land scarcity. Under high population density, the opportunity cost of land is expected to rise, reflecting increasing shortages

Figure 12.4. Per-hectare present value (discount rate = 10 percent) of increment system compared with a three-year cropping in a six-year-cycle bush fallow system in southwestern Nigeria

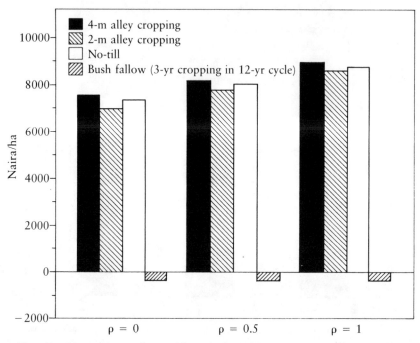

Note: Penalties are imposed on yields in the bush fallow systems in proportion to land occupied by the bush fallow, reflecting the case of high population density. ρ denotes the expected percentage of change in terminal net worth to be received by the farmers.

of arable land. To account for this, maize yields in shifting-cultivation systems are discounted in proportion to total land out of production (i.e., fallowed land). Thus, for the twelve-year-cycle bush fallow system, a 75 percent yield reduction is postulated because about three-quarters of the land is fallow. In the case of the six-year-cycle bush fallow system, a 50 percent reduction in yield is assumed because about 50 percent of the land is out of production. As expected, these yield penalties reduce the current profit advantage of the twelve-year-cycle bush fallow system as the future yield damages from these systems are increased. The long-run yield gains from topsoil conservation with the twelve-year-cycle bush fallow are now insufficient to offset the double handicap of higher costs and direct yield penalties within the twenty-year time horizon.

The 4-m alley-cropping system now yields the highest incremental return with a PVINR of N7376.5/ha. The no-till and 2-m alley-cropping systems follow with PVINRs of N7333.6 and N6994.9/ha. As expected, the twelve-year-cycle bush fallow system yields the lowest incremental return with a PVINR of −N351.4/ha. It is worth noting that under this high-population-density scenario, the traditional bush fallow systems are actually not profitable. The calculated net present values are negative and estimated to be −N2371.8/ha and −N2723.2/ha for the twelve-year- and six-year-cycle bush fallow systems, respectively. Thus, as land value rises, the returns to investment in soil-conserving systems (especially the 4-m alley-cropping and no-till systems) increase. Therefore, we expect the 4-m alley-cropping system or the no-till to be most attractive in relatively high-density population areas.

To evaluate the effects of farmers' different valuations of future incomes on the profitability of alternative fallow management systems, one conducts sensitivity analysis with respect to a higher discount rate (35 percent). A discount rate higher than 10 percent is chosen, reflecting a higher opportunity cost of capital due to the severe credit limit and the subsistence nature of small-scale farming in southwestern Nigeria. Note that in general, discount rates are a measure of the relative value of current versus future returns. For example, if the discount rate is zero, future incomes are valued equally to current incomes. If the discount rate is positive, future returns are valued less than the current ones. Thus the higher (lower) the discount rate is, the less (more) future returns are valued relative to current ones.

Results in terms of the present value of incremental net returns are reported in Figure 12.5 for the low- and high-population-density scenarios discussed above. As expected, the higher discount rate tends to postpone the profitability of soil-conserving fallow management systems. In scenario 1 (where no yield penalties are imposed in the bush fallow systems due to low population density), the twelve-year-cycle shifting-cultivation system still yields the best return, with a profitability premium (PVINR) of N823.1/ha. Clearly, when land is abundant, the twelve-year-cycle bush fallow system with nine years of fallow remains the most attractive from a profitability viewpoint. However, the no-till now ranks second with a PVINR of N575.7. The 4-m alley-cropping is next, with a PVINR of N232.1/ha. The 2-m alley-cropping system is last, with a PVINR of −N54.1/ha. It is thus

Figure 12.5. Impacts of higher discount rate (35 percent) on the per-hectare present value of incremental net returns of four alternative land use systems compared with a three-year cropping in a six-year-cycle bush fallow system in southwestern Nigeria

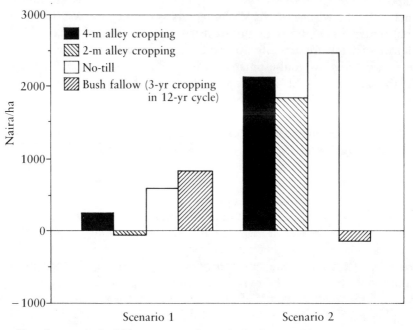

Note: In scenario 1, yields are expressed over the land occupied by the maize crop only. Scenario 2 represents the case of high population density (penalties are imposed on yields in the bush fallow systems in proportion to land in fallow).

less profitable than the more erosive six-year-cycle bush fallow system.

In the second scenario (where penalties are imposed on maize yields in traditional bush fallow systems), the no-till system gives the best incremental returns with a PVINR of N2471.3/ha. The alley-cropping system with 4-m interhedgerow spacing now ranks second, with a PVINR of N2128.0/ha, followed by the 2-m alley-cropping system, which yields a PVINR of N1841.6. The twelve-year-cycle bush fallow system ranks last, with a PVINR of −N131.9. Not surprisingly, both traditional bush fallow systems are found to be actually nonprofitable under this high-population-density scenario. Net present values in this case are estimated to be −N780.3/ha and −N912.2/ha for the twelve-year- and six-year-cycle bush fallow systems, respectively.

These results confirm Boserup's (1981) hypothesis that there exists a positive correlation between intensity of land use and population density. For given agroclimatic conditions, increases in population density will gradually move the agricultural system from forest fallow to annual cultivation. Thus, intensive cultivation of permanent fields in the frontier becomes the norm only when arable land is exhausted (Pingali and Binswanger, 1987). That the no-till system becomes more profitable than 4-m alley cropping under a high-discount-rate scenario can be attributed to the establishment cost of the alley-cropping system. It takes up to two years before any benefit can be expected from the shrubs (Kang et al., 1985; Mustaers and Spencer, 1988).

Note also that projected standard yields under the no-till system have been used despite evidence by agronomists of lower than standard yields due to increased disease, weed control, and germination problems under the seedbeds (Harder et al., 1980). It can therefore be concluded that where land value rises as a result of land shortages, farmers with lower discount rates are likely to adopt the 4-m alley-cropping system over the no-till system. For those farmers exhibiting high discount rates under high population density, research should focus on improving current profitability of the 4-m alley-cropping system.

Conclusions

Soil erosion and related environmental problems in the humid and subhumid tropics of Africa are a cause for concern. Most soils of these regions are susceptible to a rapid decline in soil productivity. Using a capital-budgeting approach, this study evaluates the profitability of alternative soil management technologies, taking into account the short-and long-run productivity effects of soil erosion in southwestern Nigeria. The technologies include two continuous-cultivation alley-cropping systems with leucaena hedgerows planted at 2-m and 4-m intervals, the continuous-cultivation no-till farming system, and two traditional bush fallow systems with a three-year cropping period in six- and twelve-year cycles. Data were collected from southwestern Nigeria, where the most common soil categories are alfisols. These soils have shallow rooting depth and are vulnerable to compaction and erosion.

Using the three-year cropping in six-year-cycle bush fallow system as the control, we compute present values of incremental net returns under two scenarios. A discount rate of 10 percent is used for the base case, which reflects the opportunity cost of capital for Nigeria. In scenario 1, where no yield penalty is imposed (reflecting the case of low population density), the twelve-year-cycle bush fallow is the most profitable, with a PVINR of N2546.8/ha (assuming the farmer receives nothing beyond twenty years of returns from the change in salvage value, due to the new technology). It is followed by the 4-m alley-cropping, the no-till, and the 2-m alley-cropping systems, with PVINRs of N1755.0, N1712.1, and N1373.3/ha, respectively. We conclude that in general, where access to new forest lands is "cost-less," slight yield damage from erosion will not detract significantly from the immediate profit advantage of traditional bush fallow systems with long fallow periods.

In the second scenario, the profitability of alternative fallow management systems is examined, taking into account total land use (cultivated and fallow). In this case, alley cropping with 4-m interhedge-row spacing is the most profitable system, with a PVINR of N7376.5/ha. No-till and 2-m alley cropping are next, with PVINRs of N7333.6 and N6994.9/ha, respectively. Not surprisingly, the twelve-year-cycle bush fallow system yields the lowest incremental return, with a PVINR of −N351.4/ha.

These findings are consistent with Boserup's hypothesis that a positive correlation exists between population density and agricultural intensification. In land-abundant areas, the fertility of the soil is maintained by periodic fallowing of land, and bush fallow systems are cost-effective. Only when farming intensifies do labor-demanding continuous-cultivation systems such as 4-m alley-cropping become profitable. Thus, where land value rises owing to population pressures (with current four- to six-year-cycle bush fallow systems), farmers have been forgoing long-term revenues in pursuit of short-term gains. Investment in soil-conserving systems such as 4-m alley cropping or no-till yields relatively higher returns, taking into account the declining yield effect of soil erosion. From a private point of view, knowledge of the cost of erosion damage may therefore be useful to farmers interested in adopting soil-conserving systems.

Sensitivity analysis was conducted to see how changes in the discount rates affect the profitability of alternative fallow management systems. Not surprisingly, a higher discount rate of 35 percent (reflecting high opportunity cost of capital) tends to postpone the profitability of alternative soil-conserving systems in favor of the traditional (more erosive) bush fallow system with 50 percent farming intensity. However, where land is abundant, the traditional bush fallow system with 25 percent farming intensity remains the most profitable, followed immediately by the no-till system. Under high-population-density conditions, the no-till system is most cost-effective, followed by the 4-m alley-cropping system. This occurs mainly because of the establishment cost of the alley-cropping system. Perhaps the most important result of this study is that it confirms Pingali et al.'s (1987) hypothesis that in land-abundant countries, strategies based on area-expansion bush fallow systems with long fallow periods are the lowest-cost sources of growth. Soil management technologies that require substantially higher labor input (e.g., the 4-m alley-cropping systems) or respond more to extra inputs (e.g., no-till) become cost-effective as population densities increase.

Before closing, it is useful to qualify the conclusions obtained by reviewing the limitations of the study. Although it yields some insight about the productivity impacts of soil erosion, this study does not provide sufficient information regarding the potential for integration of alternative fallow management systems within existing farming systems. To test if the technologies fit into the farmers' production plan, one must perform economic analysis based on a whole-farm modeling approach. Whole-farm models reflect the basic production processes involved in agriculture (e.g., nitrogen-fixing capabilities of leguminous trees) as well as many of the resource characteristics and constraints with which farmers must work (labor, land, and credit, to name but a few).

Bibliography

Aneke, D. O. "Coping with Accelerated Soil Erosion in Nigeria." *Journal of Soil and Water Conservation,* 41(3):161–63, 1986.

Boserup, E. *Population and Technological Change: A Study of Long-Term Trends.* Chicago: University of Chicago Press, 1981.

Crosson, P., and A. Stout. *Productivity Effects of Cropland Erosion in the United States.* Washington, D.C.: Resources for the Future, 1983.

Ega, L. A. "Land Tenure as a Constraint on Agricultural Development in Nigeria." In E. J. Nwosu, ed., *Achieving Even Development in Nigeria: Problems and Prospects.* Enugu, Nigeria: Fourth Dimension Publishing, 1985.

Ehui, S. K., and T. W. Hertel. "Deforestation and Agricultural Productivity in the Côte d'Ivoire." *American Journal of Agricultural Economics,* 71(3):703–11, 1989.

El-Ashry, M. T., and B. J. Ram. "Sustaining Africa's Natural Resources." *Journal of Soil and Water Conservation,* 42(4):224–27.

Ervin, C., and D. Ervin. "Factors Affecting the Use of Soil Conservation Practices: Hypotheses, Evidence and Policy Implications." *Land Economics,* 59:277–92, 1982.

Ervin, M., and J. W. Mill. "Agricultural Land Markets and Soil Erosion: Policy Relevance and Conceptual Issues." *American Journal of Agricultural Economics,* 67(5):938–92, 1986.

Food and Agriculture Organization (FAO). *A Provisional Methodology for Soil Degradation Assessment.* Rome: FAO, 1979.

Greenland, D. J., and R. Lal. *Soil Conservation and Management in the Tropics.* New York: John Wiley and Sons, 1977.

Harder, F. W., C. L. Peterson, and E. Dowding. "Influence of Tillage on Soil Nutrient Availability, Physical and Chemical Properties, and Winter Wheat in Northern Idaho." Paper presented at the Tillage Symposium, Bismarck, N. D., September 9–13, 1980.

Hoekstra, D. A. "*Leucaena leucocephala* Hedgerows Intercropped with Beans: An Ex-ante Analysis of a Candidate Agro-forestry Use System for the Semiarid Areas in Machakos District, Kenya." *Agroforestry Systems,* 1:335–46, 1982.

International Bank for Reconstruction and Development (IBRD). *Nigeria Agricultural Sector Review.* Revised, September 18, 1987, AF4AG. Washington, D.C.: World Bank, 1987.

International Institute of Tropical Agriculture (IITA). *Farming Systems Program Annual Report, 1985.* Ibadan, Nigeria: IITA, 1986.

Kang, B. T., and B. S. Ghuman. "Alley-cropping as a Sustainable Crop Production System." Paper presented at the International Workshop on Conservation Farming on Hillslopes, Taichung, Taiwan, March 20–29, 1989.

Kang, B. T., H. Grimme, and T. L. Lawson. "Alley-cropping Sequentially Cropped Maize and Cowpeas with Leucaena on a Sandy Soil in Southern Nigeria." *Plant and Soil,* 85:267–77, 1985.

Lal, R. *Soil Erosion Problems on an Alfisol in Western Nigeria and Their Control.* IITA Monograph no. 1. Ibadan, Nigeria: International Institute of Tropical Agriculture, 1976.

——. "Soil Erosion Problems on Alfisols in Western Nigeria. VI: Effects of Erosion on Experimental Plots." *Geoderma,* 25:215–30, 1981.

——. "Soil Erosion and Its Relation to Productivity in Tropical Soils." In El-Swaify and Moldenhauer, eds., *Soil Erosion and Conservation.* Antery, Iowa: Soil Conservation Society of America, 1985.

——. "Soil Surface Management in the Tropics for Intensive Land Use High and Sustained Production." *Advances in Soil Science,* 5:1–109, 1986.

Lal, R., and B. T. Kang. "Management of Organic Matter in Soils of the Tropics and Sub-Tropics." Proceedings of the Twelfth International Soc. Soil Sci., Delhi, India, 3:152–78, 1982.

Lal, R., P. A. Sanchez, and R. W. Cummings, Jr., eds. *Land Clearing and Development in the Tropics.* Amsterdam: A. A. Balkema, 1986.

Matlon, P. J., and D. S. C. Spencer. "Increasing Food Production in Sub-Saharan Africa: Environmental Problems and Inadequate Technological Solutions." *American Journal of Agricultural Economics,* 66(5):671–76, 1984.

Mutsaers, H. J. W., A. Adeyemo, T. A. Akinlusotu, K. Cashman, E. O. Lucas, J. A. Odumbaku, A. O. Ogunkunle, O. A. Osiname, J. B. Oyedokun, and Y. A. Salawu. *An Exploratory Survey of the Ayepe On-Farm Research Pilot Area, Oyo State.* OFR Bulletin no. 4. Ibadan, Nigeria: International Institute of Tropical Agriculture, 1987.

Mutsaers, H. J. W., and D. S. C. Spencer. "On-Farm Research: A Necessary Tool in the Development of Innovations." *Entwicklung und Ländlicher Raum,* March 10–12, 1988.

Ngambeki, D. S. "Economic Evaluation of Alley-cropping Leucaena with Maize-Maize and Maize-Cowpea in Southern Nigeria." *Agricultural Systems,* 17:243–58, 1985.

Oyo North ADP. "Goals, Activities, Targets, and Achievements, January 1983–March 31, 1989." Unpublished manuscript, Oyo State, Nigeria.

Pingali, P. L., Y. Bigot, and H. P. Binswanger. *Agricultural Mechanization and the Evolution of Farming Systems in Sub-Saharan Africa.* Baltimore: John Hopkins University Press, 1987.

Raintree, J. B., and F. Turray. "Linear Programming Model of an Experimental Leucaena-Rice Alley-cropping System." *IITA Research Briefs,* 1(4):5–7, 1980.

Sabel-Khoshella, U., R. Lal, and U. Schwetmann. "Runoff and Soil Erosion Studies in the Savannah." *IITA Research Highlights,* pp. 148–49, 1984.

Sanchez, P. *Properties and Management of Soils in the Tropics.* New York: John Wiley and Sons, 1976.

Seitz, W., and E. Swanson. "Economics of Soil Conservation from the Farmer's Perspective." *American Journal of Agricultural Economics,* 62:1083–88, 1980.

Sumberg, J. E., J. McIntire, C. Okali, and A. Atta-Krah. "Economic Analysis of Alley Farming with Small Ruminants." ILCA Bulletin no. 28, September 1987.

Taylor, D. B., and D. L. Young. "The Influence of Technological Progress on the Long-Run Farm Level Economics of Soil Conservation." *Western Journal of Agricultural Economics,* 10(1):63–76, 1985.

Verinumbe, I., H. C. Knipscheer, and E. E. Enabor. "The Economic Potential of Leguminous Tree Crops in Zero Tillage Cropping in Nigeria: A Linear Programming Model." *Agroforestry Systems,* 2:129–38, 1984.

Walker, D. J. "A Damage Function to Evaluate Erosion Control Economics." *American Journal of Agricultural Economics,* 64(5):690–98, 1982.

Walker, D. J., and D. L. Young. "The Effect of Technical Progress on Erosion Damage and Economic Incentives for Soil Conservation." *Land Economics,* 62(1):82–93, 1986.

Young, A. "The Potential of Agroforestry for Soil Conservation. Part 1. Erosion Control." ICRAF Working Paper no. 42, November 1986.

Young, A. R. J. Cheatle, and P. Muroya. "The Potential of Agroforestry for Soil Conservation. Part 3. Soil Changes under Agroforestry (SCUAF): A Predictive Model." ICRAF Working Paper no. 44, March 1987.

13

Resource Degradation, Agricultural
Change, and Sustainability in Farming
Systems of Southeastern Nigeria

Abe Goldman

Introduction

The "sustainability" of agricultural production has recently become an issue of concern to national and international institutions of agricultural research and policy (CGIAR, 1988). Agricultural research centers are exhorted both to examine the sustainability of the technologies they have already developed and to devise new techniques specifically to enhance the sustainability of agricultural systems. These

This chapter is based on survey research conducted for the resource management working group of the Resource and Crop Management Program of IITA. The survey was conceived, designed, and carried out as a collaborative effort among Karen Dvořák, Joost Foppes, Jonas Chianu, and the author, all of IITA, with strong support from the Imo State Agricultural Development Project (ISADAP). The author is grateful to O. Nduaka, project manager, and E. Okoro, chief research officer, of ISADAP, whose assistance and interest helped make the project possible, and to the numerous extension agents and supervisors working with ISADAP who devoted their time to organizing the village interviews. A. N. Merengini of the Michael Okpara College of Agriculture also provided valuable advice during the course of the project. At IITA, Karen Dvořák, I. O. Akobundu, B. T. Kang, Ted Lawson, Felix Nweke, Mwenja Gichuru, and Dunstan Spencer all made valuable comments and contributions to the project and to earlier written drafts. The skills of Jonas Chianu as an interpreter have been of key importance. Finally, the author and all the collaborators are grateful to the many Imo State farmers who tolerated the long interviews and provided information about their lives, work, and problems with little prospect of direct reward.

often involve complex practices whose compatibility with existing systems is uncertain.

Concerns regarding sustainability have played important roles in the ways in which past and present farming systems have evolved and changed (Ruthenberg, 1980; Sutton, 1984). Current attempts to develop sustainable new techniques, therefore, can usefully be informed by an examination of the context of sustainability issues in agricultural societies—which issues have arisen, especially in situations of high resource pressure, how farmers have tried to address them, and the roles they have played in agricultural change. This should provide insight into factors that are instrumental in the adoption of complex technologies to enhance sustainability.

This study examines some aspects of agricultural and resource management systems in southeastern Nigeria, where the International Institute of Tropical Agriculture's (IITA) Resource and Crop Management Program recently undertook a series of research projects. Among them was a survey of resource management practices in areas with differing levels of population density but similar ecological and cultural conditions (Dvořák, 1988; Foppes, 1988; Goldman, 1988). The aim of this study is to examine the threats to sustainability in this region and the ways in which farmers have addressed them.

The Context of Sustainability Issues in Southeastern Nigeria: Population and the Resource Base in Imo State

As in much of the humid tropics in West Africa, low soil productivity is one of the principal constraints on agricultural production in southeastern Nigeria. Most of the soils in the humid forest zone in this region are heavily leached, acidic, sandy Ultisols whose productivity declines rapidly after cultivation (Lekwa, 1979; Moss, 1969; Obihara, 1961; Vine, 1954, 1956). Regeneration of fertility generally requires a fallow period of several years during which woody vegetation helps to recycle nutrients and organic matter, enhances favorable microbiological activity, inhibits weeds, and protects the soil from the high temperatures and heavy rainfall that could degrade productivity (Nye and Greenland, 1960).

These sandy Ultisols cover large portions of Imo, Rivers, and Akwa Ibom states in southeastern Nigeria. The survey reported here was conducted in the acid soil region of Imo State, the southernmost of the two Ibo states of southeastern Nigeria (Figure 13.1). In most of

Figure 13.1. Location of survey areas

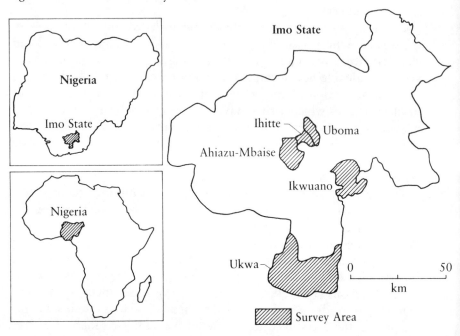

this region the major crops are cassava, yam, and oil palm, and fields are planted for only one season before being left fallow, with standing cassava often allowed to remain for about six months after harvest of the other crops.

Past fallow periods in these areas are commonly thought to have been seven years or more, often considered the minimum required for "stable yields" (Nye and Greenland, 1960; Obi and Tuley, 1973). A recent review of research on soil rest requirements concluded that at low levels of input use, Ultisols of this kind should be planted no more than one year in five to avoid degradation of productivity (Young and Wright, 1980). Even at intermediate input levels, cultivation frequency for these soils should not rise above one year in three. Survey results indicated that fallow periods in many parts of the region are already less than four years; thus, if these norms are valid, the sustainability of these agricultural systems is imperiled.[1]

---

[1] Some researchers dispute the view that fallow periods have fallen below the level required to maintain stability, noting that stable yield levels have been found in cropping rotations of much shorter duration in the area (Vine, 1954).

It is important to note, however, that the negative elements of the area's resource endowment are balanced, at least in part, by the rich and varied natural vegetation found even in areas where bush fallow systems have almost entirely replaced the forest. This vegetation and its by-products not only regenerate soil productivity for crop production but also provide numerous noncrop outputs that are of great importance to the livelihood of the people in the region (Okafor, 1981; Okere, 1983; Walker and Latzke-Begemann, 1985).

The key threat to the sustainability of agricultural systems in the region is its very large population—at density levels among the highest in sub-Saharan Africa. Imo State as a whole may have a population of around 6 million—more people than nine other countries in West Africa[2]—and a statewide density of between 430 and 520 persons per kilometer. After Lagos, it is the most densely populated of Nigeria's twenty-one states.

Because of the high population density and the resulting pressure on land and soil resources, there have been dire predictions about the likely collapse of the area's agricultural system for at least the last fifty years (Chubb, 1961; Faulkner and Mackie, 1933; Forde and Scott, 1946). One of the aims of this analysis is to explore the ways in which the farmers of this region have managed to avoid such a collapse and the lessons that can be drawn from this.

Survey Sites and Methods

The surveys reported here were conducted between November 1987 and July 1988 in five regions of Imo State (Figure 13.1). Two of these are currently local government areas (LGAs), while the other three comprise areas within LGAs that were "clan regions" or country councils under former local government administration.[3] The regions were selected on the basis of varying population density; estimates of current density (based on projections from 1953 census data) are in-

---

[2]Based on projections from the 1953 census. Accurate current figures are not available since the last accepted census was conducted in 1963, and that was suspected of significant inaccuracies (Yesufu, 1968). The other countries include Gambia, Guinea-Bissau, Mauritania, Togo, Benin, Liberia, Sierra Leone, Niger, and Senegal (World Bank, 1986).

[3]Ahaizu Mbaise (generally referred to below as Mbaise) and Ukwa are the two LGAs; Ihitte and Uboma are former clan regions currently in Etiti LGA; and Ikwuano is a section of Ikwuano/Umuahia LGA.

Table 13.1. Population densities and current and past fallow periods in the survey regions

	Estimated population density (per sq km)[b]	Past[a] (years)	Present (years)	% Change	% Cases with zero change
Ahiazu Mbaise (N = 14)	1190				
Mean		2.3	0.9	60%	7%
Range		0–5	0–2	0–100	
Ihitte (N = 9)	791				
Mean		3.5	2.2	36.1%	22%
Range		1–3	0–2	0–57	
Uboma (N = 12)	553				
Mean		4.3	3.3	19.8%	42%
Range		2–7	2–6	0–67	
Ikwuano (N = 11)	341				
Mean		5.5	4.6	15.0%	45%
Range		5–7	4–6	0–43	
Ukwa (N = 10)	170				
Mean		6.7	2.9	54.3%	0%
Range		5–9	2–4	33–78	
TOTAL SAMPLE[c] (N = 50)					
Mean	—	4.5	2.7	40.6%	17%
Range		1–9	0–6	0–100	

[a] Farmers were asked what the most common fallow period was in the village twenty years ago and what it is at present. Sample sizes refer to the number of villages sampled.

[b] Population density estimates (per sq km) are based on assumed 2.5% annual growth rate from 1953 census figures.

[c] A balanced sample of fifty villages, ten from each region, is used for the "total sample" figures in order not to bias the results by region. In the figures for each region, sample sizes vary as indicated.

cluded in Table 13.1. Both the highest- and lowest-density LGAs in the 1953 census were included in the sample (Ahiazu Mbaise and Ukwa, respectively). Earlier studies had been conducted in three of the survey regions (Morgan [1955] in Ukwa; Oluwasanmi et al. [1966] in Uboma; Lagemann [1977] in Ikwuano). The survey unit was the village. Interviews were conducted with groups of farmers who were asked to characterize conditions and practices found in their villages.

Resource Degradation and the Range of Sustainability Issues

In virtually all the villages surveyed, farmers perceived some degradation in the productivity of their resource base over the past twenty years and observed that their agricultural systems are facing a series of important issues, if not a crisis, of sustainability. Resource degradation has taken a number of forms, and the extent of degradation has not been equally severe in all areas. The main aspects of degradation are as follows:

- Decline in the yield of most arable crops. Farmers in virtually all areas say they have experienced these declines, which seem to be caused partly by declines in soil fertility and partly by increased losses due to pests and diseases.
- A restriction in the range of arable crops planted, and thus a loss of diversity in the cropping systems.
- Changes in the extent and diversity of fallow vegetation, with reduced availability of some species and a general restriction of access to resource products derived from the fallow bush.

The main immediate cause of resource degradation throughout the area is probably the large-scale reduction in fallow periods. In most of the villages surveyed, farmers felt that population pressure on available land had compelled a reduction in fallow periods, leading to declining yields and other negative effects. Table 13.1 summarizes the survey data (on a village basis) on current and past fallow periods in the five major survey regions.

Among other things, the survey found reduced fallow periods in most villages and in all of the survey regions over the last two decades or so, with a mean reduction over the total sample of about 40 percent. Current fallow periods in most places are now reported to be well below the five to seven years that may be necessary for stable yields. There is, however, extensive variation among regions and among villages in reported fallow periods and the degree to which they have changed. In some villages, fallow periods had not shortened at all over the last twenty years, while in other cases they are said to have been eliminated entirely.

In addition, the sequence by region of mean past and current fallow periods and percentage reductions follows the sequence of population densities, with the densest regions having the shortest fallow periods and greatest percentage reductions. The one anomaly is Ukwa, the area with the lowest density in Imo State in 1953. This region had the longest reported mean fallow periods in the past but has since apparently experienced an exceptionally rapid decline in fallow lengths. Present reported fallow periods there are comparable with those in regions that were much more densely populated in 1953.

These reductions in the length of fallow periods seem to be directly related to at least two of the aspects of resource degradation: crop yield declines and changes in the fallow vegetation.

Yield Declines

In virtually all of the villages surveyed, farmers said they had experienced some yield decline over the past two decades. Although it is difficult to quantify the extent of the declines, we tried to obtain rough estimates by asking farmers the number of stands of cassava (the major food crop throughout the region) needed to fill a basket now compared with the number needed twenty years ago.[4] Even in the instances of moderate yield decline, farmers estimated past yields of cassava to have been around twice those obtained at present. In severe cases, past yields are said to have been over 10 times, sometimes up to 30 times, present yields.

Farmers often ascribe these yield declines to the reduced fallow periods, showing a general correspondence with the amount by which fallow periods are said to have been reduced in the area. From farmers' responses, however, it appears that reduced soil productivity may not be the only factors involved. Yield declines were reported even in cases in which there was little or no reduction in fallow periods, and farmers in many instances said that the yields they obtained were lower than past levels even on fields fallowed for the same length of time. A possible cause could be increased losses to pests and diseases.[5]

---

[4] Note that this results in relative estimates of yield per plant, not necessarily yield per unit area. This seemed more consistent with the way farmers actually conceive of crop yields, especially in a comparative sense. However, the translation between yield per plant and yield per acre is not necessarily straightforward, as it is affected not only by stand density but, especially for cassava, by the length of time the crop is allowed to remain in the ground before harvest.

[5] For example, recent surveys of the cassava mealybug (*Phenacoccus manihoti*) in southeastern Nigeria by the Africa-wide Biological Control Program suggest that both the incidence and severity of damage caused by the mealybug increase in areas with shorter fallow

Agronomic practices that can affect yields have also changed in some areas, particularly the length of time cassava is left to mature before harvest. Farmers often attribute these changes to declines in soil fertility. In addition, yield declines have occurred mainly on farmers' "outer fields," but not necessarily on fields that, like the household compound gardens, are more intensively managed with the addition of much organic matter and nutrients. Fertilizer use has also increased substantially in at least some areas (especially in the densely populated Mbaise region), though it has only retarded, not eliminated, yield declines.

Crop Declines and Disappearances

In most villages surveyed, farmers noted that a number of crops had declined in importance or disappeared altogether over the past twenty years. In most cases the declines were not matched by the introduction of new crops, and there has been a net reduction in crop diversity. This may indicate degradation of the germ plasm resource base in the farming system. The most serious and widespread decline has occurred in the case of cocoyams (both the *Xanthosoma* and *Colocasia* species, although more severely for the former), which fifteen to twenty years ago in many areas exceeded yams in importance and were second only to cassava (Lagemann, 1977; Oluwasanmi et al., 1966).

This decline, which was cited by farmers in all of the villages surveyed, is ascribed to a radical downturn in yields, which, based on the major symptoms described and other reports (Nnoke et al., 1987), is probably the cocoyam root rot blight complex.[6] Cocoyam is still planted to some extent (often in compound gardens), especially the more resistant *Colocasia* varieties, but in general the crop has lost its former prominence, and the formerly preferred *Xanthosoma* species

---

periods, especially on the sandy, low-nutrient soils characteristic of this region (P. Neuenschwander, personal communication). Further analysis is needed to confirm this preliminary finding.

 [6] This is a soil-borne disease caused by a complex of agents, chief of which is the fungus *Pythium myriotylum* (IITA, 1981; Théberge, 1985). Also known as cocoyam declining disease, it is said to have appeared in southeastern Nigeria within the last ten years and is estimated to be causing losses of up to 80 percent on *Xanthosoma sagittifolium* (Nnoke et al., 1987). It is reckoned now to be the principal limiting factor to cocoyam in Africa and threatens to become the major limiting factor throughout the humid tropics (Caveness et al., 1987).

is said to have declined sharply. The major replacement has been cassava.

In addition to cocoyam, farmers reported declines in groundnuts, lima beans, yam beans *(Sphenostylis stenocarp)*, velvet beans *(Mucuna sloanei)*, coconuts, some yam species (including aerial yam, *Dioscorea bulbifera,* and trifoliate yam, *D. dumeotrum*), and some other crops. With the exception of the minor yam species, which have declined mostly because people have lost interest in eating them, the declines in other crops are often ascribed to yield reductions due to the incidence of new or increasingly virulent diseases or pests. Moreover, at present, plantains, an important food and income crop throughout much of the area, are threatened by the recent appearance of black sigatoka disease *(Mycosphaerella fijiensis* var. *difformas),* which is likely to cause high losses and possibly a major reduction in plantain production (IITA, 1988).

These reductions in crops may diminish the farming systems' resilience in responding to other ecological or economic fluctuations. In addition, many of the losses have been of locally cultivated legumes which have not been replaced. This may have had a negative effect on the overall nitrogen balance of the cropping system, since few annual nitrogen-fixing species remain. The reduction in plantain production would also reduce an important semiperennial species. Both of these trends can have negative impacts on soil productivity as well as reducing species diversity.

There have been few new crops in the region to balance these losses. Those that have been introduced (rice, pineapples, citrus, etc.) have generally been confined to certain areas, although the Imo State Agricultural Development Project (ISADAP) is now attempting to introduce new cassava varieties and other crops over the entire region. In the early 1960s, the Uboma Project attempted to introduce and/or expand a range of new crops and techniques in the Uboma region: wetland rice, hybrid maize, improved cassava varieties, vegetables, citrus, pineapples, and improved oil palms, as well as the use of fertilizers and Aldrin dust as seed dressing for yams (Oluwasanmi et al., 1966). Twenty years later, although some may have been adopted, the only crop to have had a significant and lasting impact has been wetland rice, which requires swamp drainage and bund construction (supported by the project). Although highly productive, these drained areas are limited in size and have not spread beyond the areas of their initial subsidized introduction, either within or outside Uboma.

Degradation of Fallow Vegetation and Related Resources

Although primary and most secondary forest has been eliminated in most of the survey region, there remains a varied range of bush vegetation that regrows rapidly once fields are left uncultivated. Interspersed among it are a few older forest trees that have been preserved because of their economic importance. This fallow vegetation is the key to the regeneration of soil productivity after the cultivation period, which is why the reduction of fallow periods is so critical.

Beyond its ecological function in regenerating soil productivity, fallow vegetation provides a range of resource products, including firewood, stakes for yams and other crops, food, construction materials, and medicinal raw materials. As fallow periods have shortened, the amount and quality of these resource products have diminished, and farmers now frequently use inferior species for many of these functions or have to purchase products that were previously common property resources. Resource pressures are not equally severe in all areas, but shortages in some areas can help to induce commodification of resource products and restriction of access rights even in areas where there is no internal shortage.

Our survey focused mainly on fallow vegetation used for fuelwood and yam stakes,[7] including the main species farmers used for their purposes, how they have changed over time, and the extent to which they have become commodities bought and sold within or outside the village. There appear to be five good indicators of the extent of resource pressure on these fallow products:

- Whether staking material is purchased from sources outside the village
- Whether fuelwood is purchased from sources outside the village
- Whether bamboo is one of the main species used for yam stakes
- Whether palm fronds are one of the main species used for fuelwood
- Whether and to what extent access is restricted to these raw materials in the village's bushland

Table 13.2 summarizes the percentage of villages in each survey region reporting these conditions.

---

[7] Yam vines have to be staked in this environment in order to grow properly, and farmers may need from several hundred to several thousand yam stakes per season. Depending on the method used, the type of yam, and the desired tuber size, stakes can range from moderate-sized tree branches or split bamboo poles to heavy poles that may be three or more meters tall.

Table 13.2. Indicators of resource pressure on fallow vegetation, by survey region (percentage of positive responses)

Survey region	Stake purchases[a]	Fuelwood purchases[b]	Bamboo as stakes[c]	Palm fronds as fuelwood[d]	Access restrictions Stakes	Access restrictions Fuelwood
Mbaise (N = 14)	92% (13)	29%	93%	85% (13)	100%	36%
Ihitte (N = 10)	90%	67% (9)	100%	44% (9)	86% (7)	0% (8)
Uboma (N = 13)	22% (11)	0%	69%	0%	62%	22% (9)
Ikwuano (N = 11)	9%	0%	9%	0%	9%	0%
Ukwa (N = 10)	30%	30%	40%	0%	50%	10%

Note: The numbers in parentheses represent varying sample sizes for specific responses.
[a] Purchasing of poles for yam stakes from sources outside the village.
[b] Purchasing of fuelwood from sources outside the village.
[c] Bamboo as the first or second most used species for yam stakes in the village.
[d] Palm fronds as the first or second most used species for fuelwood in the village.

Stake and/or fuelwood purchases outside the village are likely to indicate shortages within the village. Stems of the bush *Acioa barteri* make the best staking material because of their strength and termite resistance. *Acioa* is widely available in most areas, partly because it was planted for this purpose in the past. Use of bamboo for yam stakes has begun relatively recently. Bamboo stakes are said to be acceptable, but they do not last as long as *Acioa* and they often must be purchased since bamboo grows only in certain locations. Palm fronds make poor fuel but are freely available; their widespread use for fuel probably indicates serious shortages of other fuelwood species.

In general, the data show much the same interregional pattern that emerges from the data on fallow periods: there is a decline in the degree of pressure on fallow vegetation resources moving from Mbaise through Ihitte, Uboma, and Ikwuano corresponding with 1953 population densities, but Ukwa is an anomaly in relation to its low density in the past.[8]

The low rate of fuelwood purchases in Mbaise compared with Ihitte and even Ukwa, in contrast to other indications of high pressure, is somewhat easier to account for. As discussed below, although the quantity of fallow vegetation in Mbaise is generally limited, farmers

[8] The reason for Ukwa's anomalous position in terms of both fallow length and resource pressure is unclear. However, a key factor may be that out-migration from the area seems to be lower and to have started later than in the other regions (Goldman, in press).

there practice an intensive agroforestry system based on *Acioa,* which supplies large amounts of staking and fuelwood material. This illustrates the need to consider farmers' responses to resource pressure and to examine the degree to which their responses have been successful in retarding degradation.

Resource Management Practices and Responses to Resource Degradation

Various elements of the farming systems in Imo State can be considered resource management practices. These range from discrete individual practices, such as fertilizer use, to relatively complex techniques that make up subsystems within the farming system as a whole. They help to mitigate the impact of reduced fallow periods and deal with the main aspects of resource degradation. The measures are often multipurpose. Resource management may not have been their main original purpose, nor may it be their only function at present. Nevertheless, they play this role in the farming systems and are often viewed by farmers as such.

The resource management systems and practices discussed below include:

- Fertilizer use
- The maintenance of high-intensity production niches, including compound gardens and wetland rice areas
- Indigenous agroforestry systems
- Land-use transfers

Table 13.3 summarizes some aspects of these resource management systems, including their major functions, costs and limitations, and general regional prevalence. These are examined in more detail below.

Fertilizer Use

Use of commercial fertilizers seems the most straightforward way of obtaining higher yields and overcoming at least some of the problems caused by shorter fallow periods.[9] Fertilizer trials at the National

---

[9] However, organic matter additions are necessary in these soils in addition to fertilizer, partly to help combat the acidification that can be caused by nitrogenous fertilizers (Kang and Juo, 1983).

Measure and system	Function and advantage	Costs and trade-offs	Prevalence and distribution by region
High-intensity pro-duction niches			
Compound farms	Garden crops and staples continuously planted; maintained by inputs of animal manure, household wastes, etc.; helps overcome problems of seasonality; crop diversity; yield declines; food security	High labor input; limited materials availability	Present in all areas; in high-density areas, management is more intensive and compounds may be proportionately larger and more important
Wetland rice areas	Rice planted annually; numerous other crops planted on residual moisture; extensive fertilizer use and other intensive management; has opened new land for cultivation; increased food and income output; introduced new crop(s) and preserved some old crops	High initial development costs; high labor input; materials costs	Present only in Uboma because of prior development project there
Indigenous agro-forestry systems	Certain species, especially *Acioa barteri,* planted or protected in fallow fields; rapid regrowth during fallow period; supplies yam stakes and firewood; enhancement of soil regeneration; other species provide semi-wild food products	Roots may compete with crops; delayed benefits to current planting; soil improvement of *Acioa* not as good as other species	Some past *Acioa* planting in most areas; intensive current planting only in highest-density area
Fertilizer use	Increases yields; permits use of land with short fallow (partial substitute for fallow regeneration)	Cash for purchase (subsidized price); said to shorten storability of yam and cassava	Low usage in most low- and medium-density areas; high use only in highest-density areas
Land-use transfers and resource product sales	Farmers obtain farming land on short-term basis from others in same or different community to help overcome land shortage and/or effects of short fallow periods; diffuses some impacts of fallow reduction	Cost of land rental of fallow products; distance and money constraints; possible increased fallow depletion	Occurs in almost all areas; probably highest in lower-density areas close to cities or high-density areas

Table 13.4. Reported fertilizer adoption

	Estimated % fertilizer adoption within village	
	≥40%	<40%
Mbaise (N = 14)	93	7
Ihitte (N = 10)	0	100
Uboma (N = 13)	31	69
Ikwuano (N = 11)	18	82
Ukwa (N = 10)	10	90
TOTAL (N = 58)	32	68

Root Crops Research Institute (NRCRI, formerly known as the Federal Agricultural Research and Training Station) at Umudike (located in our survey region in Ikwuano) have consistently shown positive responses in cassava yields from the addition of nitrogen, phosphorus, and potassium, and somewhat lower responses to calcium and magnesium (Federal Agricultural Research and Training Station, 1975, 1976). For the last several years, fertilizers have been available at subsidized prices in Imo State through the ISADAP distribution centers. The use of fertilizers is widely promoted by local extension agents and demonstrated on numerous small "corner plots" throughout the state.

Nonetheless, the results of our survey suggest that fertilizer use remains low in most areas. The reasons cited by farmers include lack of knowledge (which fertilizers to use, how to apply them, etc.), limited cash availability and high costs, and a widespread belief that fertilizers reduce the storability of yams. The latter may be the major obstacle because the belief is pervasive, even among farmers who have not themselves tried fertilizers. Experiments at NRCRI show no direct evidence of this, although scientists there suggest that the increased water content of fertilized yam tubers may make them bruise more easily when harvested and therefore make them more susceptible to rot (E. Nnodu, personal communication). It is also suggested that relatively less nitrogen and more potassium would be more appropriate for yam nutrition than the currently used 15:15:15 fertilizer and should reduce the storage problem.

In our survey, farmers were asked how many in the village were using fertilizers. The results are summarized in Table 13.4.

In only about one-third of the villages in the total sample were 40 percent or more of the farmers using fertilizers. In general, fertilizer adoption seems to vary with population density, with one exception. In the highest-density area, Mbaise, virtually all villages reported high levels of fertilizer use. In contrast to other areas, farmers in Mbaise said that the reduced storability or quality of yam and cassava tubers due to fertilizer use was either not significant or was overridden by the necessity of using fertilizers. Fertilizer use is extremely low in Ihitte, although this is also an area of very high density where fallow periods have been shortened and yield levels are said to have declined sharply. In Uboma[10] and Ikwuano, fertilizer adoption is intermediate, while in Ukwa, it is very low or nonexistent in most villages.

On the whole, fertilizer adoption throughout the region (with the exception of the Mbaise area) is lower than expected, given the various factors that would seem to encourage its use: yield-increasing response, relatively low subsidized price, wide availability, and extensive promotion. More study is needed to determine the reasons for these relatively low rates of adoption and what might lead to increased use.

Even if the obstacles were removed, however, the experience in Mbaise suggests that increased fertilizer use may make only a limited contribution to overcoming the yield decline problem and enhancing the sustainability of agricultural production. Although farmers in Mbaise said that fertilizer use there began in the 1970s, sharp yield declines were reported over this period despite the uncharacteristically high use of fertilizer.

Compound Gardens as High-Intensity Production Niches

In contrast to the usual cultivation cycle of a single planting season followed by a fallow period of varying duration, certain fields are planted on a continuous basis. These are always managed intensively, with use of various fertility-maintaining inputs to overcome the sharp productivity decline that would otherwise occur. These "high-intensity production niches" make up distinct components of the farming systems. There are two main examples in the survey areas: the compound gardens maintained by most households, and a small number of areas where wetland rice cultivation has been developed.

One major way of mitigating the effects of shortened fallow periods is to manage the nutrient-recycling process, increasing its efficiency relative to the natural regrowth and decay of fallow vegetation. This requires considerable input of labor, compost, and manure, which are

---

[10] Much of the fertilizer use in Uboma is said to be concentrated in the intensively farmed rice areas, which are generally planted every year. These are discussed below as special production niches.

in limited supply. In the survey area (and in much of southeastern Nigeria), these recycling techniques are largely confined to compound gardens—an important part of farming systems in this region (Green, 1941; Forde and Scott, 1946).

These are small farms or gardens surrounding the household compound in which various vegetables, fruits, tree crops, and some staples are planted, usually continuously without any intervening fallow period. This is made possible by the constant addition of organic matter and other nutrients such as animal manure, food by-products, household wastes, ashes, and leaves and other plant materials. Such organic materials can stabilize soil productivity and prevent yield declines in these kinds of soils (Nye and Greenland, 1960; Vine, 1954, 1956). In addition to the direct soil additions, the profusion of trees and crops in such gardens results in a multistoried farming ecology which is able to mobilize nutrients, water, and sunlight from different levels of the soil and atmosphere (Niñez, 1987).

Such household farms are found in many, though not all, parts of the humid forest zone in West Africa, as well as in numerous other areas of the humid tropics (Fernandes et al., 1984; Okafor and Fernandes, 1987; Snelder, 1987). The system probably antedates contemporary conditions of high density and likely was originally a cultural response to various needs related to household food supply (e.g., seasonality and variety) rather than a response to land-use pressures.[11] Over time, however, the functions served by compound farms have changed, particularly with the increasing population and land pressures.

Among the main benefits of compound farms cited by farmers are that they provide foodstuffs at times of the year when other foods are not available (especially the "hungry season" before the harvest of the main staple crops [Okere, 1983; Walker, 1985]), and they provide foods, especially fruits and vegetables, that are not planted elsewhere. Further benefits include the convenience of food growing close to the home as well as the extra security for valuable foods such as yams that may be threatened by animal pests or theft. Compound gardens serve as sites for experimentation with new crops (Richards, 1985) and also help to preserve some of the crops that have been in decline. Farmers in many areas mentioned that cocoyam, yam beans, and a

[11] Among the evidence suggesting this kind of cultural linkage rather than an origin based on increasing population density is the fact that compound farms are present in both lower- and higher-density areas in southeastern Nigeria, while they are rarely found in either low- or high-density areas in southwestern Nigeria (Okafor and Fernandes, 1987).

number of the other declining crops are still being planted in the compound farms, largely by women farmers for home consumption. Perhaps higher fertility in the compounds helps them resist the diseases or other pressures that have caused the reported yield declines in the outer fields.

What is not clear is the extent to which compound farming has changed in response to population growth and how it differs among higher- and lower-density areas. There is little direct information on the nature and rate of change in any one location, but there have been some comparative cross-section studies in this region in villages of differing densities.

In his study of three Imo State villages with differing population densities, Lagemann (1977) found that compound farms were larger, supported higher crop densities, and accounted for a greater proportion of total farm output in the high-density as compared with the medium-density villages. Yields, however, were lower because comparable amounts of organic matter additions had to be dispersed over a greater area. Snelder's study (1987) of species composition on compound farms in three villages with varying densities in Rivers State did not find the correlation between density and the size of compound farms that Lagemann observed. She did, however, find that on compounds in the high-density village more annual food crops such as cassava were planted, resulting in reduced species diversity, fewer tree crops, and a diminished multistory structure than in the less densely populated villages.

In our survey, important differences were found between the intensity of compound farm use and management in the lowest-density area, Ukwa, and higher-density areas. In Mbaise, for example, all households have compound farms, which are planted every year. Manure and compost additions are spread through the entire field, and some households also add fertilizer. In contrast, not all households in Ukwa maintain compound fields, and rather than being planted continuously, only a portion may be planted each year. In addition, manure and compost are not necessarily spread through the entire field but are often dumped in one spot, and there is no fertilizer use.

It appears likely that an increase in relative size and an intensification of nutrient recycling on the compound farms have been among the important ways in which farmers in high-density areas have managed to cope with land shortages. This is likely to occur also in lower-density areas, such as Ukwa and Ikwuano, as land pressures there increase. There may also be a blending of near fields and compound farms in areas such as Mbaise, with an extension of nutrient-recycling techniques into the former. However, farmers in most villages said that few of the nutrient-recycling techniques used in the compound farms are extended to outer fields, mainly because there is not enough compost or manure available and the outer fields are too distant to carry the material. Thus, it seems that constraints on land, labor, and materials limit both the extent of this avenue of intensification and the degree to which it can overcome the problems of yield declines.

There are no obvious prospects for either of these constraints to be overcome. Animal manure in particular is limited because there are no cattle in the area, only goats, sheep, and poultry. Moreover, in most of the villages surveyed, farmers said that the numbers of goats and sheep have declined because they now have to be tethered to keep them out of other peoples' fields. In Mbaise, there seems to be an extension of compost use from the compounds into the nearer fields, and the two farming areas may have begun to combine, but this is largely a result of shrinking farm sizes, and not a solution to declining yields and output. As fertilizers are more extensively used in the high-density areas, they may also be applied on the compound gardens, especially as these expand. The character of the compound gardens will probably also change in the process, with more staple crops planted there and perhaps a reduction in the overall diversity of crop species (Snelder, 1987). These changes, together with the need to diffuse inputs and management over larger areas, may lead to some yield decline in the compounds and the outer fields. Again, as with fertilizer use, intensification through compound gardens may retard the long-term decline in yields but is not likely to reverse it.

Wetland Rice Development

The other major example of a special production niche is of much more recent origin than the compound farms. Wetland rice cultivation has been introduced into many parts of southeastern Nigeria over the last forty years, mostly in the plateau areas with hydromorphic soils, and to a lesser degree in flooded riverine areas and swamplands. The introduction of rice has often transformed the farming systems and the structure of land use in these areas (Agboola, 1979). In the region of our survey, however, the impact of wet rice production has been much more restricted overall, though where it has been introduced, rice cultivation has had a dramatic impact.

Rice is grown to a significant extent in only one of the five areas we surveyed. It was introduced into Uboma as part of the Shell Uboma Project in the mid-1960s (Oluwasanmi et al., 1966). By 1972, the area under rice cultivation in Uboma was estimated at about 430 ha (Anthonio and Ijere, 1973). Substantial capital and technical assistance were provided by the project and the Ministry of Agriculture for dam construction and swamp and paddy development, as well as for initial seed supply and training. In the Uboma villages where rice is currently grown (seven of thirteen survey villages), farmers now consider it one of the top four or five food crops (much of it is grown for sale); in some cases, rice is rated second only to cassava.

Perhaps even more important, this has resulted in the creation of an entirely new, intensively managed production niche. In most cases, rice is planted every year in the wet season, and a variety of vegetables and other crops (including cocoyam and some maize and cassava) are planted as dry-season crops. Not only do many households have small individual plots in the rice areas, but farmers from neighboring villages who want to plant rice can rent plots of land from the owners. Moreover, fertilizer use is widespread in these areas, and mulching and other soil and water management techniques are common. Essentially, these irrigated rice areas have become a major locus for intensive arable-crop production.

Many villages, of course, do not have areas suitable for rice or have not been able to develop the areas that are suitable. Indeed, wetland rice development in some villages has not stimulated similar development in the other Uboma villages—people still try to rent rice plots on an individual basis. Neither within nor outside Uboma did we encounter a case of a village or group of villages that had organized on their own and invested the capital and labor needed for rice land development. This suggests either that the returns on such an investment may not be adequate to justify the costs involved without some subsidy or that the level of organization required may be too great an obstacle. However, this experience does demonstrate the potential for creation of a major new niche for intensive production.

Indigenous Agroforestry Systems

If the composition of fallow vegetation can be managed to produce greater amounts of organic matter and other soil nutrients in a shorter time, then the efficiency of soil fertility restoration can be enhanced, and shorter fallow periods may be as effective as longer ones.[12] In much of Imo State and some other parts of southeastern Nigeria, an indigenous agroforestry system has been practiced for a long time. In most areas this involves planting or tending the fast-growing bush species *Acioa barteri,* whose stumps regenerate rapidly after coppicing and which has hard, termite-resistant wood with a number of uses. The existence of this "planted fallow" system has been noted by past researchers (Faulkner and Mackie, 1933; Morgan, 1955), and there have been experimental studies of *Acioa* fallows on soil quality (Nye and Hutton, 1957; Obihara, 1965; Vine, 1954). There have also been attempts, some going back over forty years, to extend *Acioa* planting to other areas for erosion control as well as for fertility enhancement, though these have not been successful (Chubb, 1961; Grove 1951a, 1951b; Obihara, 1965).

Acioa is an important fallow species in most of the villages in our survey; farmers said it had been deliberately planted by their forefathers, probably at least eighty years ago. In some cases *Acioa* seed is still planted, mainly to fill in gaps or occasionally in fields in which it does not occur. Most of the *Acioa,* however, is in the form of old stumps with extensive root systems from which the bushes regenerate rapidly after the crops are removed. The stumps are fire-resistant and survive field burning (they are also partially protected). The branches are removed during field clearing and are used for yam stakes, firewood, and construction.

Although many farmers say that *Acioa* is among the best fallow species for improving soil fertility (probably at least partially because of its rapid regrowth and extensive leaf production), it was originally planted more for high-quality yam stakes than for soil improvement. Farmers in a number of places feel that *Acioa* is not as good for improving soil fertility as are a number of other species, and some farmers even feel it is detrimental, largely because of its root system. It is the quality of wood that is the motivation for the deliberate planting and encouragement of *Acioa.*

In our survey regions, *Acioa* is present or is purposively managed to differing degrees, but the most intensive management undoubtedly occurs in Mbaise, the highest-density area. Farmers there report that

---

[12] This is the basis of the alley-cropping system, which essentially combines the fallow and cropping periods by planting crops between rows of rapidly growing leguminous trees whose trimmings are left on the ground to be incorporated into the soil (Kang and Wilson, 1987).

because of the very short fallow periods, the range of natural fallow species remaining is extremely limited, but most maintain one or more fields in which *Acioa* has been densely planted—usually in rows—and is allowed a longer fallow time in order to produce large enough yam stakes. (After two seasons or more of use, the stakes are burned for fuel; they are one of the main fuel sources in the region.) In other words, this is a system of field differentiation, producing a "wood crop" between periods of food-crop production. In Mbaise, this has clearly helped farmers to cope with the pressure on fallow vegetation resources.

The origin of the system dates back before the current intense land-use pressures and cannot be said to be a direct result of those pressures. Moreover, in the nearby area of Ihitte, which is also very densely populated, there is no comparable intensive agroforestry system. Indeed, farmers there said that *Acioa* has declined markedly as fallow periods have shortened. It is not clear why an intensive *Acioa* system did not take hold in Ihitte or why it has not been adopted recently as land-use pressures have intensified. This example illustrates that increasing density does not alone ensure development or adoption of intensive management systems that may be available and beneficial.

Land-Use Transfers

Many who have written about this area in the past have commented on the extensive movement of farmers within the region to obtain land for farming (Chubb, 1961; Forde and Scott, 1946; Green, 1941; Harris, 1942). Transfers of land-use rights are usually on a short-term basis, either as gifts or for rent, or are pledges in which money is loaned in return for use of the land until repayment (Mbagwu, 1978). Taken as a whole, this amounts to a regionwide system of interchanges of land-use rights in which land shortages in some areas (or in individual households) are mitigated by the use of land in other areas in which supply is more plentiful. In general, the system is supported by strong social norms, and farmers feel it is appropriate to supply land to those who need it, especially if there is any relationship of kinship or friendship. Even in some of the most densely populated areas, outsiders come to obtain land for farming, which they are often given.

All of our survey villages participate in one way or another in land interchange. Some are primarily "exporters" of land (i.e., outsiders come there but most villagers there do not go elsewhere to obtain land); others are primarily "importers" (village members mostly go out, but few come in for land). There are also numerous cases in which both export and import of land use occurs. Both the amount of land available in the village and the proximity to an area with a differing level of land availability help determine the extent of interchange in any one place. The details of these patterns are discussed more fully elsewhere (Goldman, in press), but it is clear that this remains an important social means of diffusing the impact of land-use pressures and ameliorating localized degradation.

Conclusions

Reviewing the situation of the regions in Imo State, we find a number of threats to the resource base that may jeopardize the sustainability of agricultural systems. Farmers reported various forms of resource degradation over the past twenty years in most, though not all, places. Declines in soil productivity, in the quality, diversity, and availability of fallow vegetation, and in the diversity of planted crop species are cited by farmers in most areas. The consequences have included reduced crop yields, perhaps increased pest and disease pressures, increased food purchases, increased commodification and decreased access to the resource products deriving from fallow vegetation, and the loss of the consumption values and contributions to system stability and resilience provided by some of the crops that have declined. Shortened fallow periods play a key role in this process, especially in this agricultural ecology where the fallow period is essential to the soil-vegetation system.

This gloomy depiction of a system caught in an irreversible and interlinked downward spiral is, however, modified by at least two sets of factors. First, the extent of resource pressure varies sharply within the region, and generalizations about the state of degradation do not hold for all portions of it; and, second, farmers have devised and adapted a variety of means of coping with and retarding resource declines, ranging from discrete practices to entire subsystems in which intensive resource management and input use are concentrated in particular sites.

These responses have had varying degrees of success, and some are found only in scattered locations, but even in the region where resource pressure is highest, one or more aspects of degradation have been ameliorated or arrested. In addition, there are widespread social mechanisms for the interchange of productive resources, including particularly the provision of land for short-term use through renting, gift, or other means to those from areas in which it is in short supply. This is a means of making use of the diversity in states of resource pressure among different regions, and it has for a long time diffused the impact of threats to the local resource base.

All of these means of ameliorating and coping with the hazards of resource degradation involve serious costs or limitations. To some extent, they can probably continue to expand to areas in which they are not yet widely used (e.g., there seems to be much potential for expanded and improved fertilizer use), or they can be intensified and improved in areas where they already occur. They will probably, however, have to be supplemented by "new" techniques and systems. In trying to develop and extend these to farmers, the characteristics of the prevailing resource management systems and the changes that were successfully introduced in the past should be kept in mind.

Many of the resource management practices discussed here did not originate primarily as such, at least not in the sense of trying to cope with resource pressures and degradation. In particular, the compound farming system, the *Acioa*-based agroforestry system, and even the extensive land-use transfer practices originated long before contemporary conditions of ultrahigh population densities and widespread resource degradation. They seem, however, to have been adapted as a means of dealing with these kinds of conditions. Elements of the systems have also occasionally been split off and applied in other locations or systems (as when some of the nutrient-recycling techniques of the compound farms are applied in wetland rice areas). This suggests that many of the changes to come in these farming systems will also originate as adaptations and extensions of existing practices rather than as wholly novel transplants. Consequently, agricultural research, especially for resource management technology, might be most successful through a strategy of adaptation and "rearrangement" of elements that already exist rather than through devising entirely new packages.

Closely related to this is another characteristic common to all of the resource management systems discussed: in all of them, the "resource management" components are closely linked to increases in short-term productivity and output of crops or other agricultural commodities. Planting of *Acioa*, which so much resembles current alley-cropping systems, was devised and is now practiced principally to provide yam stakes, not for soil improvement, although many farmers believe it does accelerate regeneration of soil fertility. There is no comparable widespread system of tree planting solely for soil improvement, though this would also increase crop yields.

Similarly, compound farming and wetland rice systems, which include an ensemble of soil management practices, are intended specifically to increase production of certain desired crops, for cash income as well as for home consumption. Systems of *Mucuna*-based mulching or *Acioa*-based tree planting purely for soil management have a history of failure in southeastern Nigeria (Chubb, 1961; Forde and Scott, 1946). The lesson here is that, particularly for complex systems of resource management, innovations or adaptations are most likely to be taken up when they are linked to systems that increase production of desired commodities. This should be incorporated into the research and development strategies for resource management techniques.

Related to this is the converse: new "productive technology" inherently includes resource management implications. The introduction of new higher-yielding, disease-resistant cassava varieties in this region should help relieve some pressure on the soil resource base. Ideally, in order to achieve greatest impact, such varieties should be bred for the range of resource conditions that prevail in this area—that is, short fallow periods and minimal fertilizer use in most cases (but in some cases higher fertilizer use and/or longer fallows). Since it is likely that the variation currently found in this area will continue into the future (though its specific character may change), there should be a range of improved crop varieties for each of the major types of conditions. Among the most beneficial of innovations would be the introduction of highly productive and profitable new subsystems, such as the wetland rice system, that employ underutilized resources or ecological niches and incorporate sets of resource management techniques— preferably adapted from existing practices—to ensure sustainable and profitable production.

Finally, although it is easy to be pessimistic about an area as crowded as this but not apparently well endowed with productive resources, any dire predictions should be tempered by the realization that the imminent collapse of the system has been assumed numerous times over at least the last half century. The system has not collapsed over this time, though it has had to change, and there have certainly been costs in the form of resource degradation. Most likely, a "collapse" of the region and system is not imminent, but further changes must occur, many of them impelled by the need to cope with and perhaps reverse the declines that have occurred in the productivity of the resource base.

Bibliography

Agboola, S. A. *An Agricultural Atlas of Nigeria*. Oxford: Oxford University Press, 1979.

Anthonio, Q. B. O., and M. O. Ijere. *Uboma Development Project, 1964–1972*. London: Shell International, 1973.

Caveness, F. E., S. K. Hahn, M. N. Alvarez, and Y. Ng. "The Cocoyam Improvement Program at IITA, 1973 to 1987." In O. B. Arene et al., eds., *Cocoyams in Nigeria: Production, Storage, Processing and Utilization*. Umudike, Nigeria: National Root Crops Research Institute, 1987.

CGIAR. "Sustainable Agricultural Production: Implications for International Agricultural Research." Rome: TAC Secretariat and FAO, 1988.

Chubb, L. T. *Ibo Land Tenure*. 2d ed. 1947. Ibadan: Ibadan University Press, 1961.

Dvořák, K. A. "Resource Management in the Humid Tropics: Working Paper on Defining Research Domains." Unpublished manuscript, IITA, 1988.

FAO. *Atlas of African Agriculture*. African Agriculture: The Next 25 Years. Rome: FAO, 1986.

Faulkner, O. T., and J. R. Mackie. *West African Agriculture*. London: Cambridge University Press, 1933.

Federal Agricultural Research and Training Station [Umudike]. *Annual Reports*. Umudike, Nigeria: Federal Agricultural Research and Training Station, 1975, 1976.

Fernandes, E. C. M., A. Oktingati, and J. A. Maghembe. "The Chagga Home Gardens: A Multi-storeyed Agro-forestry Cropping System on Mt. Kilimanjaro, Northern Tanzania." *Agroforestry Systems*, 2:73–86, 1984.

Floyd, Barry. *Eastern Nigeria: A Geographical Review*. London: Macmillan, 1969.

Foppes, J. "Survey on Fallow Management Practices in Bori LGA, Rivers State, Nigeria." Unpublished draft report, IITA, 1988.

Forde, Daryll, and Richenda Scott. *The Native Economies of Nigeria*. London: Faber and Faber, 1946.

Goldman, Abe. "Population Growth and Agricultural Change in Imo State, Southeastern Nigeria." In R. W. Kates, G. Hyden, and B. L. Turner, eds., *Population Growth and Agricultural Intensification: Studies from Densely Settled Areas of Sub-Saharan Africa*. Gainesville: University of Florida Press, in press.

Green, M. M. *Land Tenure in an Ibo Village in South-Eastern Nigeria*. Monographs on Social Anthropology, no. 6. London: Percy Lund, Humphries, for the London School of Economics and Political Science, 1941.

Grove, A. T. *Land Use and Soil Conservation in Parts of Onitsha and Owerri Provinces*. Geological Survey of Nigeria, Bulletin no. 21. Lagos: Government Printer, 1951a.

———. "Soil Erosion and Population Problems in Southeast Nigeria." *Geographical Journal,* 117:291–306, 1951b.

Harris, Jack. "Human Relationships to the Land in Southern Nigeria." *Rural Sociology,* 7:89–92, 1942.

IITA. *Research Highlights.* Ibadan, Nigeria: IITA, 1981.

———. *IITA Annual Report and Research Highlights, 1987/88.* Ibadan, Nigeria: IITA, 1988.

Kang, B. T., and A. S. R. Juo. "Management of Low Activity Clay Soils in Tropical Africa for Food Crop Production." In *Proceedings of the Fourth International Soil Classification Workshop,* Rwanda, June 2–12, 1981. Brussels: ABOS, AGCD, 1983.

Kang, B. T., and G. F. Wilson. "The Development of Alley Cropping as a Promising Agroforestry Technology." In H. A. Steppler and P. K. R. Nair, eds., *Agroforestry: A Decade of Development.* Nairobi: ICRAF, 1987.

Lagemann, Johannes. *Traditional African Farming Systems in Eastern Nigeria: An Analysis of Reaction to Increasing Population Pressure.* Munich: Weltforum Verlag, 1977.

Lekwa, Godwill. "The Characteristics and Classification of Genetic Sequences of Soils in the Coastal Plain Sands of Eastern Nigeria." Ph.D. diss., Michigan State University, 1979.

Mbagwu, T. C. "Land Concentration around a Few Individuals in Ibgo-Land of Eastern Nigeria." *Africa,* 48:101–15, 1978.

Morgan, W. B. "Farming Practice, Settlement Pattern, and Population Density in Southeastern Nigeria." *Geographical Journal,* 121:322–33, 1955.

Moss, R. P. "An Ecological Approach to the Study of Soils and Land Use in the Forest Zone of Nigeria." In M. F. Thomas and G. W. Whittington, eds., *Environment and Land Use in Africa.* London: Methuen, 1969.

Niñez, V. "Household Gardens: Theoretical and Policy Considerations." *Agricultural Systems,* 23:167–86, 1987.

Nnoke, F. N., O. B. Arene, and A. C. Ohiri. "Effect of N.P.K. Fertilizer on Cocoyam Declining Disease Control and Yield in *Xanthosoma sagittifolium.*" In O. B. Arene et al., eds., *Cocoyams in Nigeria: Production, Storage, Processing and Utilization.* Umudike, Nigeria: National Root Crops Research Institute, 1987.

Nye, P. H., and D. J. Greenland. *The Soil under Shifting Cultivation.* Technical Communication no. 51. Harpenden: Commonwealth Bureau of Soils, 1960.

Nye, P. H., and R. J. Hutton. "Some Preliminary Analyses of Fallows and Cover Crops at the West African Institute for Oil Palm Research, Benin." *Journal of the West African Institute for Oil Palm Research,* 2(7):237–43, 1957.

Obi, J. K., and P. Tuley. "The Bush Fallow and Ley Farming in the Oil Palm Belt of Southeastern Nigeria." Miscellaneous Report 161, Land Resources Division, Ministry of Overseas Development (ODM), U.K., 1973.

Obihara, C. H. "The Acid Sands of Eastern Nigeria." *Nigerian Scientist,* 1(57):57–64, 1961.

———. "Effect of *Acioa barteri* Fallows on the Fertility of an Acid Sandy Soil in Nigeria." In *Proceedings of the OAU/STRC Symposium on the Maintenance and Improvement of Soil Fertility,* Khartoum, November 8–12, 1965. London: OAU Publications Bureau, publication 98, 1965.

Okafor, J. C. "Woody Plants of Nutritional Importance in Traditional Farming Systems of the Nigerian Humid Tropics." Ph.D. diss., University of Ibadan, Nigeria, 1981.

Okafor, J. C., and E. C. M. Fernandes. "Compound Farms of Southeastern Nigeria." *Agroforestry Systems,* 5:153–68, 1987.

Okere, L. C. *The Anthropology of Food in Rural Igboland.* Lanham, Md.: University Press of America, 1983.

Oluswasanmi, H. A., et al. *Uboma: A Socio-economic and Nutritional Survey of a Rural Community in Eastern Nigeria.* World Land Use Survey, Occasional Papers, no. 6. Bude, Cornwall, England: Geographical Publications Limited, 1966.

Richards, Paul. *Indigenous Agricultural Revolution: Ecology and Food Production in West Africa.* London: Hutchinson; Boulder, Colo.: Westview Press, 1985.

Ruthenberg, Hans. *Farming Systems in the Tropics.* 3d ed. Oxford: Clarendon Press, 1980.

Snelder, D. "A Case Study on Compound Farms in Southeastern Nigeria." RCMP Project Report. Ibadan: IITA, 1987.

Sutton, J. E. G. "Irrigation and Soil Conservation in African Agricultural History." *Journal of African History,* 25:25–41, 1984.

Théberge, Robert L., ed. *Common African Pests and Diseases of Cassava, Yam, Sweet Potato and Cocoyam.* Ibadan, Nigeria: IITA, 1985.

Udo, R. K. "Patterns of Population Distribution and Settlement in Eastern Nigeria." *Nigerian Geographical Journal,* 6:73–88.

Vine, H. "Is the Lack of Fertility of Tropical African Soils Exaggerated?" *Proceedings of the Second Inter-African Soils Conference,* Leopoldville, 1954.

———. "Studies of Soil Profiles at the WAIFOR Main Station and at Some Other Sites of Oil Palm Experiments." *Journal of the West African Institute for Oil Palm Research,* 1(4):8–59, 1956.

Walker, Judith. "Interim Report: Compound Farming Systems of Southeastern Nigeria." Unpublished report. Ibadan: IITA, 1985.

Walker, Judith, and Ute Latzke-Begemann. "List of Plant Species and Their Uses in Southeastern Nigeria." Unpublished report. Ibadan: IITA, 1985.

World Bank. *World Development Report, 1985.* New York: Oxford University Press, 1986.

Yesufu, T. M. "The Politics and Economics of Nigeria's Population Census." In J. C. Caldwell and C. Okonjo, eds., *The Population of Tropical Africa.* London: Longmanns, 1968.

Young, A., and A. C. S. Wright. "Rest Period Requirements of Tropical and Subtropical Soils under Annual Crops. In *Report on the Second FAO/UNFPA Consultation on Land Resources of Populations of the Future.* Rome: FAO, 1980.

Index

Library of Congress Cataloging-in-Publication Data

Moock, Joyce Lewinger.
  Diversity, farmer knowledge, and sustainabilty / Joyce Lewinger Moock, Robert E. Rhoades.
    p. cm.—(Food systems and agrarian change)
  Includes bibliographical references and index.
  ISBN 0-8014-2682-0.—ISBN 0-8014-9968-2 (pbk.)
   1. Sustainable agriculture—Congresses. 2. Agricultural diversification—Congresses.
3. Agriculture—Research—Congresses. 4. Agricultural systems—Congresses. 5. Agriculture—
Technology transfer—Congresses. I. Rhoades, Robert E. II. Title. III. Title: Farmer
knowledge. IV. Series.
S494.5.S86M66 1992
338.1'6'091724—dc20   92-52768